"中国森林生物多样性监测网络"丛书　　马克平 主编

北京东灵山暖温带森林
——物种组成与分布格局

Beijing Donglingshan Warm Temperate Forest Dynamics Plot: Species Composition and Their Spatial Distribution Patterns

祝　燕　林秦文　王兴煜　邓婷婷　姚红霞　刘　博　杜晴晴　著

中国林业出版社
China Forestry Publishing House

图书在版编目 (CIP)数据

北京东灵山暖温带森林：物种组成与分布格局 / 祝燕等著 . —北京：中国林业出版社，2023.12

ISBN 978-7-5219-2532-6

Ⅰ. ①北… Ⅱ. ①祝… Ⅲ. ①暖温带 - 森林植物 - 研究 - 北京 Ⅳ. ①S718.3

中国国家版本馆CIP数据核字(2024)第007437号

内容简介

本书详尽地描述了北京东灵山暖温带落叶阔叶林常见木本植物49种，每个树种除了文字描述外，还配有精美的照片，展示植物的小枝、花序、果实和幼苗等，方便识别。同时，附有每种植物在20 hm² 长期定位研究样地内的种群分布图以及种群的个体数量和径级结构，对于该样地的地形、土壤、植被类型等皆有介绍。本书以资料翔实、图片精美为特色，暖温带落叶阔叶林研究不可多得的参考书，也可以作为植物爱好者了解暖温带森林、认识森林植物的野外指导手册。

中国林业出版社

策划、责任编辑：于界芬
电话：010-83143542

出　　版：中国林业出版社 (100009　北京西城区德内大街刘海胡同7号)
网　　址：http://lycb.forestry.gov.cn/
发　　行：中国林业出版社
印　　刷：河北京平诚乾印刷有限公司
版　　次：2023年12月第1版
印　　次：2023年12月第1次
开　　本：889mm×1194mm　1 / 16
印　　张：8.25
字　　数：210千字
定　　价：108.00元

序 言 1

在过去的几十年时间里，中国科学院和林业、农业等相关部门陆续建立了数百个生态系统定位研究站。其中，中国科学院组建的中国生态系统研究网络 (CERN) 拥有分布于全国包括农田、森林、草地、湿地、荒漠等生态系统类型的36个生态站。国家林业和草原局建立的中国森林生态系统研究网络 (CFERN) 由29个生态站组成，基本覆盖了我国典型的地带性森林生态系统类型和最主要的次生林、人工林类型。

随着研究的发展，特别是近年来人们对生物多样性和全球变化研究的关注，国际上正在推动生态系统综合研究网络平台的建立。在全球水平上，全球生物多样性综合观测网络 (GEO BON) 是一个有代表性的研究网络。它试图把全球与生态系统和生物多样性长期定位研究相关的网络整合起来，通过综合研究，探讨生态系统与生物多样性维持与变化机制以及系统之间的相互作用机理，为生态系统可持续管理与生物多样性的保护提供科学依据和管理模式。

近年来，中国科学院生物多样性委员会组织建立了中国森林生物多样性监测网络 (Chinese Forest Biodiversity Monitoring Network，以下简称CForBio)。中国是生物多样性特别丰富的少数国家之一，也是唯一一个具有从北部寒温带到南部热带完整气候带谱的国家。截至2018年，中国森林生物多样性监测网络包括大型监测样地17个，成为继美国史密森研究院热带研究所建立的全球森林生物多样性监测网络 (CTFS-Forest GEO) 之后又一大型区域监测网络。由于CForBio横跨多个纬度梯度，对于揭示中国森林生物多样性形成和维持机制，以及森林生物多样性对全球变化的响应，科学利用和有效保护中国森林生物多样性资源具有重要意义。

目前，CForBio已经有很好的研究进展，各样地研究成果陆续在国际著名生态学刊物如 *Ecology Letters, Journal of Ecology* 等上发表，受到国内外同行的高度评价。但这些文章都是关于某一具体问题的研究总结，还无法让国内外同行全面了解CForBio各个样地整体情况。因此，出版这套以中英文形式介绍各大样地基本情况的"中国森林生物多样性监测网络"丛书是非常必要的。感谢马克平研究员组织相关专家编写这套丛书。我相信该丛书不仅是国内外同行深入了解CForBio各样地的参考书，同时也将为我国森林生物多样性监测和森林生态系统联网研究奠定重要的基础。

（孙鸿烈）

中国科学院前副院长

Foreword 1

In the past few decades, hundreds of Ecosystem Research Stations have been set up by the Chinese Academy of Sciences, State Forestry Administration, Ministry of Agriculture and other relative departments. Among them, 36 ecological research stations were established by Chinese Ecosystem Research Network (CERN), supported by the Chinese Academy of Sciences. The 36 research stations are scattered over the country representing diverse ecosystems, including farmland, forest, grassland, wetland, desert and others. Moreover, the Chinese National Ecological Research Network (CFERN), supported by the State Forestry and Grassland Administration, consists of 29 research stations, covering typical zonal forest ecosystems and main secondary forests and plantations in China.

With the development of research, especially the growing concern over researches on biodiversity and global change in recent years, the establishment of ecosystem research network have been promoted under international supports. So the Group on Earth Observations Biodiversity Observation Network (GEO-BON) is representative across the world, and it attempts to integrate worldwide networks relating to long-term research on ecosystem and biodiversity. Based on the comprehensive studies, the maintenance and change mechanism of ecosystem and biodiversity and their interactions have been explored, which provide scientific basis and management mode for sustainable development of ecosystem and protection of biodiversity.

In recent years, Chinese Forest Biodiversity Monitoring Network (CFor Bio) has been built by Biodiversity Committee of the Chinese Academy of Science. China is one of the few top "mega-biodiversity countries" in the world, and it is also the only country with full climatic zone spectrum, ranging from nerthern cool temperate zone to soouthern tropical zone. Till 2018, CForBio consists of 17 large scale monitoring plots, and it is the second large scale regional monitoring network after ForestGEO organized by the Smithsonian Tropical Research Institute.

Due to its covering several latitude gradient, CForBio is significant for revealing formation and maintenance mechanisms of Chinese forest biodiversity, responses of biodiversity to global change, scientific utilization and effective protection of Chinese forest biodiversity resources.

Encouraging progress has been made in this area since the network built, for lots of research findings have been published in the international peer reviewed ecological journals, such as *Ecology Letters, Journal of Ecology* and *Oikos,* etc., which brought about positive response from colleagues in the field of plant ecology. However, the published papers mostly focus on research of specific problems; scientists and public still can't understand the whole situation of each plot in details. So it is really necessary to publish this series, which introduce basic information of permanent forest plots in both Chinese and English. I am grateful to Professor Keping Ma for organizing related specialists to prepare the series. And I believe that this series would be a valuable reference book for scientists and public to further understand CForBio, and it will also lay a foundation for the forest biodiversity monitoring and forest ecosystem research in China.

Honglie Sun

The former Vice-President for the Chinese Academy of Sciences

序 言 2

　　森林在维持世界气候与水文循环中起着根本性的作用。森林是极为丰富多样的动物、植物与微生物的家园，而人类正是依靠这些生物获取各种产品，包括食品与药物。尽管对人类福祉如此重要，森林仍然遭受着来自土地利用与全球气候环境变化的巨大威胁。在这种不断变化的情况下，为了更好地管理全球剩余的森林，迫切需要树种在生长、死亡与更新方面的详细信息。

　　中国森林生物多样性监测网络 (CForBio) 正在中国沿着纬度与环境梯度建立大尺度森林监测样地。通过这个重要的全国行动倡议与来自中国科学院及若干其他单位的研究者的努力，CForBio开始搜集关于中国森林的结构与动态的关键信息。现在CForBio与史密森研究院及哈佛大学阿诺德树木园的全球森林监测网络 (ForestGEO) 形成了合作伙伴。ForestGEO是个在27个热带或温带国家拥有长期大尺度森林动态研究样地的全球性网络。CForBio与ForestGEO合作的目标是通过合作研究，了解森林是如何运作的，它们是如何随着时间而改变的，以及如何重建或者恢复，以确保森林提供的环境服务能可持续或者增长。森林及其提供的服务的长期可持续性有赖于我们预测森林对全球变化，包括气候与土地利用变化的响应的能力，以及我们去理解与创建适当的森林服务市场的能力。通过拥有67个森林大样地的全球网络及大量项目的训练与能力建设，CForBio与ForestGEO的伙伴关系是发展这些预测工具的重要基础。这种伙伴关系也将促进为全球各地的当地社区、林业管理者与政策制定者在森林的保育与管理方面发展应用性的林业项目建议，发展与示范利用乡土物种进行森林重建的方法，以及从经济学角度评估森林在减缓气候变化、生物多样性保护和流域保护上的价值的方法。

　　我祝贺作者们创作了这部关于样地植物的优秀丛书。本丛书为将来的森林监测提供了基准信息，是涉及森林恢复、碳存储、动植物关系、遗传多样性、气候变化、局地与区域保育等研究内容的研究者、学生与森林管理者们有价值的参考资料。

S.J. 戴维斯

主任

史密森热带研究所全球森林监测网络

Foreword 2

Forests play an essential role in regulating of world's climatic and hydrological cycles. They are home to a vast array of animal, plant and microorganism species on which humans depend for many products, including food and medicines. Despite the importance of forests to human welfare they are under enormous threat from changes in land-use and global climatic conditions. In order to better manage the world's remaining forests under these changing conditions detailed information on the dynamics of growth, mortality and recruitment of tree species is urgently needed.

The Chinese Forest Biodiversity Monitoring Network (CForBio) that aims to establish large-scale forest monitoring plots across latitudinal and environmental gradients in China. Through this important national initiative, researchers from the Chinese Academy of Sciences and several other research institutions in China, CForBio has begun to gather key information on the structure and dynamics of China's forests. The CForBio initiative is now partnering with the Forest Global Earth Observatory (ForestGEO) of the Smithsonian Research Institute and the Arnold Arboretum of Harvard University. ForestGEO is a global program of long-term large-scale forest dynamics plots in 27 tropical and temperate countries/areas. The goal of the partnership between CForBio and ForestGEO is to work together to understand how forests work, how they are changing over time, and how they can be re-created or restored to ensure that the environmental services provided by forests are sustained or increased. The long-term sustainability of forests and the services they provide depend on our ability to predict forest responses to global changes, including changes in climate and land-use, and our ability to understand and create appropriate markets for forest services. The CForBio-ForestGEO partnership is ideally poised to develop these predictive tools through a global network of 67 large forest plots and an extensive program of training and capacity building. The partnership will also lead to the development of applied forestry programs that advise local communities, forest managers and policy makers around the world on conservation and management of forests, to develop and demonstrate methods of native species reforestation, and to economically value the roles that forests play in climate mitigation, biodiversity conservation, and watershed protection.

I congratulate the authors on the production of this excellent new series of stand books. In addition to providing a baseline for future forest monitoring, these books provide a valuable resource for researchers, students, and forest managers dealing with issues of forest restoration, carbon storage, plant-animal interactions, genetic diversity, climate change, and local and regional conservation issues.

Stuart Davies

Director
Forest Global Earth Observatory / ForestGEO
The Smithsonian Tropical Research Insitute

前 言

中国暖温带落叶阔叶林区域位于北纬32°30′～42°30′、东经105°30′～124°30′的范围，以栎类为优势的落叶阔叶林是华北山区的地带性顶极植物群落，具有较丰富的生物多样性、复杂的组成结构和较高的生态系统服务功能，是华北平原重要的生态屏障。由于长期的人为干扰，生态系统破坏十分严重，原始林几乎丧失殆尽。其中东灵山地区地处北京市门头沟区的西部山区，分布有暖温带落叶阔叶林的地带性植被类型—辽东栎林。由于人为活动极为频繁，东灵山在海拔1000 m以下至平原的低地丘陵，森林几乎荡然无存，大面积覆盖的为灌丛。天然森林主要分布在东灵山海拔1000 m以上。北京百花山国家级自然保护区1985年建立，东灵山森林植被得到较好的保护。在北京东灵山暖温带落叶阔叶林开展森林生态学研究，对暖温带植被保护、水土保持与生态文明建设都有重要意义，也将为京津冀生态环境建设发挥支撑作用，特别是将贡献于北京地区生态安全屏障建设。

为深入开展暖温带落叶阔叶林区域生态学研究，1990年中国科学院植物研究所陈灵芝研究员及其团队，在东灵山创建北京森林生态系统定位研究站，开展了三十多年的生态系统定位研究，积累了大量的资料，在暖温带森林生态系统结构与功能研究方面取得丰富的成果，为东灵山生物多样性和生态系统服务研究打下良好基础。然而，在森林生态系统维持机制方面研究相对薄弱。因此，有必要在具有代表性的暖温带落叶阔叶林中建立大型森林动态样地，长期定位监测从植物开花、结实、种子扩散、幼苗萌发直至植株建成、死亡等完整的生活史过程，同时，结合对不同营养级的监测，比如微生物、昆虫、鸟类、大型动物等，探讨森林群落构建和维持机制，以期为东灵山生物多样性保护和生态环境建设提供理论支持。

参照全球森林动态样地的建立规程，2009年11月至2010年9月选择以辽东栎为优势种的暖温带落叶阔叶林在东灵山建立了一个20 hm²的森林动态样地。目前为止，已经进行了12年的森林生物多样性监测与研究。北京东灵山森林动态样地的建成不仅对北京市及华北地区生态环境建设有重要意义，而且填补了中国森林生物多样性监测网络（Chinese Forest Biodiveristy Network, CForBio）和全球森林监测网络（The Forest Global Earth Observatory, ForestGEO）在暖温带落叶阔叶林区域的空白，为相关科研院所和高校提供了暖温带落叶阔叶林生物多样性和生态系统研究的重要平台，也是森林生态科学知识教育和传播的重要窗口。

本书详尽地描述了东灵山20 hm²暖温性落叶阔叶林样地群落的物种特性、分布格局和不同物种在样地内的径级结构等。书中精美、分辨率高的样地植物照片让读者对样地植物物种鉴定有更感性的认知。同时本书也是东灵山样地建设启动12年来工作积累的展现，为未来样地的深入研究提供必备的基础生态信息，并服务于更多的研究者、植物爱好者和自然保护工作者。

本书记录了东灵山森林动态样地木本植物49种，隶属于23科33属；草本植物87种，隶属于33科61属。本书裸子植物按照郑万钧1975年系统、被子植物按照APG IV分类系统排列。植物的生物学特性描述主要参考《中国植物志》和*Flora of China*。

东灵山森林大样地的建立和维护，是一项庞大和艰巨的事业。在样地建设及12年来的各项科研活动过程中，得到了各单位老师、研究生和诸多朋友的大力支持。借此机会，诚恳地感谢对样地建设做出贡献的所有人，特别感谢北京森林生态系统定位研究站提供了食宿等便利条件，感谢北京百花山国家级自然保护区小龙门管理站（小龙门林场）和清水镇政府等的多方面支持。

首先感谢桑卫国教授、朱丽博士分别于2010年、2015年组织完成第一次、第二次森林动态样地调查。特别感谢刘海丰博士、王顺忠博士、李亮和李文超等，和我从2009年秋冬到2010年，参加了样地的选址、建立和完成第一次调查；感谢河南科技大学、平顶山学院、河南农业大学、曲阜师范大学等高校的同学们参与由我负责的第三次植被调查；感谢王兴煜、邓婷婷、黄琛豪、姚红霞、于涛、徐帅伟和杜晴晴等几位研究生在样地复查及随后的野外工作和数据核查等方面的贡献；感谢林秦文高级工程师等为本书提供了大量专业的植物图片，感谢刘博博士在样地调查中对疑难树种和草本植物的鉴定和书籍的专业校对；感谢北京林业大学侯继华教授和张乃莉青年研究员对样地建设的支持！感谢北京森林生态系统定位研究站前站长张齐兵研究员、王杨执行站长等提供的便利；感谢小龙门林场王葛场长、汪经理；感谢王金增和郑宝胜等北京市清水镇和河北涿鹿县的师傅们！

感谢马克平研究员、中国科学院生物多样性委员会办公室徐学红博士对书稿内容的认真审校！书中还可能有错误疏漏之处，请各位读者指正。

祝燕

2022年9月20日

Preface

The warm temperate deciduous broad-leaved forest area in China is located in the range of 32°30′N–42°30′N north latitude and 105°30′E–124°30′E longitude. The deciduous broad-leaved forests dominant by oak trees with rich biodiversity, complex composition and high ecosystem services, are the zonal climax plant communities in the mountains of North China and important ecological barriers for the North China Plain. The economic development in this region was relatively early in the Chinese history, and the ecosystem was severely damaged. Due to long-term human disturbance, the primary forest was almost completely lost. Among them, the Donglingshan area is located in the western mountainous area of Mentougou District of Beijing, where there is a zonal vegetation type of warm temperate deciduous broad-leaved forest-Liaodong oak forests. Due to the extremely frequent human activities, the original forests of Donglingshan were cut down in the lowland hills below 1000m above sea level to the plains, where a large area was covered with shrubs. Natural forests are mainly distributed in Donglingshan above 1000m above sea level. Beijing Baihuashan National Nature Reserve was established here in 1985, so the vegetation of Donglingshan is well protected. Therefore, the forest ecology research in the warm temperate deciduous broad-leaved forest in Beijing Donglingshan is of great significance for the protection of warm temperate vegetation, soil and water conservation and the construction of ecological civilization, meanwhile, will play a key role in the construction of the Beijing-Tianjin-Hebei ecological environment, especially it will contribute to the construction of ecological barriers for Beijing.

In order to carry out regional ecological research on deciduous broad-leaved forests in warm temperate zones, in 1990, Prof. Chen Lingzhi and her team from the Institute of Botany, Chinese Academy of Sciences, established the Beijing Forest Ecosystem Research Station in Donglingshan, conducted many studies, accumulated a large amount of data and made rich achievements on the structure and function of warm temperate forest ecosystems. It provided a well foundation for the researchs on biodiversity and ecosystem function of Donglingshan. However, these studies were difficult to reflect many processes of forest ecosystems, and cannot deeply understand the maintenance mechanisms of biodiversity at the forest community level. Therefore, it is necessary to establish large-scale forest plots in typical warm temperate deciduous broad-leaved forests to long-term monitor the complete life history process of plants from flowering, fruiting, seed dispersal, seedling germination to establishment and death. Together with the monitoring at different trophic levels, such as microorganisms, insects, birds, large animals, etc., to explore the factors that affect the dynamics and interaction changes of plants and other multi-trophics of biodiversity, it will provide theoretical support for protecting the biodiversity and ecological environment construction of Donglingshan.

Referred to the procedure for establishing global forest dynamics plots, we established a 20 hm² permanent monitoring plot dominant by *Quercus wutaishansea* in warm temperate deciduous broad-leaved forest during the period of November 2009 to September 2010. So far, we have monitored biodiversity and conducted researches in the Donglignshan forest dynamics plot (Donglignshan FDP) for 12 years. The establishment of Donglingshan FDP is not only of great significance to the ecological environment construction in Beijing and North China, but also fills the gap within the Chinese Forest Biodiversity Network (CForBio) and the Forest Global Earth Observatory (ForestGEO) in the warm temperate deciduous broad-leaved forest area, providing an important platform for research on biodiversity and ecosystems of warm temperate deciduous broad-leaved forests for scientific research institutes and universities, and an important window for the education and dissemination of forest ecological science knowledge.

This book describes in detail the species characteristics, distribution patterns and size-class distribution of different species in the warm temperate deciduous broad-leaved forest community in the 20-hm^2 plot of Donglingshan (Donglingshan FDP). The exquisite and highly recognizable plant pictures in the plot allow readers to have a more perceptual understanding of the species identification in this book. At the same time, this book is also a demonstration of the work accumulation in the 12 years since the construction of the Donglingshan FDP. It provides necessary basic ecological information for the deep studies in the future, and attracts more researchers, plant enthusiasts, and ecological protection volunteers to join for exploring forest ecology.

This book records 49 species of woody plants belonging to 23 families and 33 genera; 87 species of herbaceous plants belonging to 33 families and 61 genera. In this book, gymnosperms are sorted according to Zheng Wanjun's 1975 system, and angiosperms are sorted according to APG IV classification system. The description of the biological characteristics of plants mainly refers to Flora of China in Chinese and English.

The establishment and maintenance of the Donglingshan plot is a huge and arduous undertaking. During the construction of the plot and various scientific research activities over the past 12 years, it has received strong support from colleagues, graduates and friends. I would like to take this opportunity to sincerely thank all for your contribution to the construction of the plot, especially thanks to the Beijing Forest Ecosystem Research Station for providing accommodation and other convenient conditions, thanks to Xiaolongmen Forest Farm in Beijing Baihuashan National Nature Reserve and the Government of Qingshui Town for their support!

First, thank Professor Sang Weiguo, Dr. Zhu Li for completing the first and second vegetation surveys in 2010 and 2015 respectively. Especially, thank Dr. Liu Haifeng, Dr. Wang Shunzhong, Li Liang, Li Wenchao and so on worked with me on the establishment of plot and the first vegetation survey; thanks to the students from Henan University of Science and Technology, Pingdingshan University, Henan Agricultural University and other colleges and universities in the third vegetation survey in 2021; thanks to Wang Xingyu, Deng Tingting, Huang Chenhao, Yao Hongxia, Yu Tao, Xu Shuaiwei and so on for contributing to the third survey and subsequent field work, clean data, etc. Thanks to Senior Engineer Lin Qinwen and others for providing professional plant pictures for this book, and to Dr. Liu Bo for plant identification on difficult identified species and herbs in the vegetation survey and book edition; thanks to Associated Professor Hou Jihua and Zhang Naili of Beijing Forestry University for their support for the construction of the plot! Thanks to the previous Director of Prof. Zhang Qibing, Dr. Wang Yang and so on in Beijing Forest Ecosystem Research Station for the convenience provided, thanks to the Director Wang Ge of Xiaolongmen Forest Farm, Manager Wang; thanks to Wang Jinzeng, Zheng Baosheng and other local residents in Qingshui Town, Beijing and Zhuolu County, Hebei! Due to space limitations, sorry for omissions!

Thank you to Prof. Ma Ping, Dr. Xu Xuehong of the Office of the Biodiversity Committee of the Chinese Academy of Sciences, for their careful review of the content of the manuscript! Due to the rush of time and the limited knowledge, mistakes and omissions are inevitable. We will appreciate if readers help to correct the book. Thank you.

Zhu Yan
September 20, 2022

目 录

Contents

东灵山地区自然植被简介
Introduction to Natural Vegetation of Donglingshan

I

1.1 地理位置和自然环境

东灵山森林样地属于北京市西部门头沟区东灵山地，离北京市区大概120 km。东灵山地是小五台山向东延伸的支脉，属太行山系，与著名的百花山山体相连。最高峰海拔达2303 m，是北京市境内的最高峰，也是华北地区的主要高峰之一。

该地区属暖温带半湿润季风气候，四季分明。据多年气象资料显示，年平均气温4.8℃，年均最高气温为18.3℃，年均最低气温-10.1℃。年降水量为588 mm，6~8月的降水量约占全年降水量的74%，7月多暴雨。

该地区地势险峻，土壤大致有褐土、棕壤、亚高山草甸土等几种类型，海拔高度及由它决定的生物气候特点、地形和地质因素在土壤形成中具有显著的作用。山地顶部为亚高山草甸土（海拔>1800 m），中山地带为山地棕壤（海拔1000~1800 m），低山地带为褐色土（海拔<1000 m）。该区土壤肥沃，土层厚度约为30 cm，富含有机质。全区土壤pH值大多介于6.0~7.5之间。

1.1 Location and Natural Environment

Donglingshan forest plot belongs to Donglingshan or known as Mount Dongling. Donglingshan is a mountain in western hills located in Mentougou District in the west of Beijing, about 120 kilometers away from the downtown Beijing, China. Donglingshan is the eastward branch of Xiaowutai Mountain. It is an extension of the Taihang mountain connected with the famous Baihua Mountain. The highest elevation is 2303 m. The summit is not only the highest peak within the Municipality of Beijing, but also one of the main peaks in North China.

The region has a warm temperate semi-humid monsoon climate with distinct seasons. According to the meteorological data in this area for many years, the annual average temperature is 4.8℃, the annual average maximum temperature is 18.3℃, and the annual average minimum temperature is -10.1℃. The annual precipitation is 588 mm. About 74% of the annual precipitation is concentrated from June to August, and there are mostly heavy rains in July.

The terrain of this area is steep, and the soil generally includes cinnamon soil, brown soil, subalpine meadow soil and other types. The altitudes and the bioclimatic characteristics, topographic and geological factors determined by it have a significant role in soil formation. The subalpine meadow soil distributes on the top of the mountain with an altitude of more than 1800 m. The mountainous brown soil mainly distributes in the Zhongshan zone with an altitude of 1000 to 1800 m. Cinnamon soil is in the low mountain zone with the altitude of less than 1000 m. The soil in this area is fertile with a thickness of about 30 cm and rich in organic matter. The soil pH in the whole area is mostly between 6.0 and 7.5.

1.2 主要植被类型

东灵山地区海拔差异较大，导致明显不同的水热条件，从而形成了东灵山自下而上明显的环境梯度，造就了东灵山类型多样的自然植被，特别是随海拔梯度的变化，呈现出明显的垂直分布带谱，构成了独特的自然景观格局。依照垂直分布特点，分为三条带：低山落叶阔叶灌丛带（海拔400~1000 m），本带阳坡常见荆条（*Vitex negundo* var. *heterophylla*）灌丛，阴坡还包括有三裂绣线菊（*Spiraea trilobata*）灌丛；中山落叶阔叶林带（海拔1000~1900 m），本区地带性植被——暖温带落叶阔叶林于此分布；亚高山草甸带（1700~2300 m），此带原为以华北落叶松（*Larix gmelinii* var. *principis-rupprechtii*）为主的山地寒温性针叶林带，由于人为破坏，原生植被荡然无存，现分布着矮紫苞鸢尾（*Iris ruthenica* var. *nana*）+细柄薹草（*Carex capillaris*）草甸和紫苞风毛菊（*Saussurea purpurascens*）+细柄薹草草甸为主的植物群落，还有金露梅（*Dasiphora fruticosa*）、银露梅（*Dasiphora glabra*）、鬼见愁（*Caragana jubata* var. *jubata*）灌丛及白桦（*Betula platyphylla*）、硕桦（*Betula costata*）林。主要植被类型见图1-1。

1.2 Main Vegetation Types

The obvious altitude gradient in Donglingshan results in various hydrothermal conditions and environmental gradient along the altitude gradient, which provides habitats for various vegetation types, especially forms the distinct vertical vegetation zones and the unique landscape patterns. According to the characteristics of vertical distribution, it is divided into three vegetation zones: The deciduous broad-leaved shrub zone located at the low mountain with an altitude of 400 to 1000 m, where *Vitex negundo* var. *heterophylla* is common at the sunny slope of this zone and *Spiraea trilobata* is dominant at the shady slope; A deciduous broad-leaved forest belt with an altitude of 1000 to 1900 m, where the warm temperate deciduous broad-leaved forest is distributed here; At the altitude of 1700 to 2300 m, there is the subalpine meadow belt, which is originally the mountainous cold temperate coniferous forest belt dominated by North China larch *Larix gmelinii* var. *principis-rupprechtii*. Due to serious human disturbance, the original vegetation disappeared. Now there are plant communities dominated by *Iris ruthenica* and *Carex duriuscula* meadow and plant community dominant by *Saussurea purpurascens* and *Carex duriuscula*, as well as *Dasiphora fruticosa*, *Dasiphora glabra*, and *Caragana jubata* var. *jubata* shrubs and *Betula platyphylla*, *Betula costata* forests. The main vegetation types are shown in the photos below (Fig. 1-1).

亚高山草甸 (拳参) Subalpine meadow (*Polygonum bistorta*)

亚高山草甸 (地榆) Subalpine meadow (*Sanguisorba officinalis*)

亚高山草甸 (华北大黄) Subalpine meadow (*Rheum franzenbachii*)

鬼箭锦鸡儿灌丛 *Caragana jubata* shrubs

鬼箭锦鸡儿灌丛 *Caragana jubata* shrubs

金露梅灌丛 *Potentilla fruticosa* shrubs

白桦林 *Betula platyphylla* forest

糙皮桦林 *Betula utilis* forest

辽东栎为主的落叶阔叶林 Deciduous broad-leaved forest dominated by *Quercus wutaishanica*

辽东栎林 *Quercus wutaishanica* forest

落叶松林（人工） *Larix gmelinii* var. *principis - rupprechtii* plantation forest

油松林（人工） *Pinus tabuliformis* plantation

图1-1 东灵山不同植被类型 (摄影：刘永刚) Different vegetation types in Donglingshan (Photo by Liu Yonggang)

东灵山暖温带落叶阔叶林样地
Donglingshan Warm Temperate Deciduous Broad-leaved Forest Plot

2

2.1 样地基本概况

东灵山暖温带落叶阔叶林动态监测样地位于北纬39°95′，东经115°42′，北京森林生态系统定位站附近，清水镇小龙门森林公园中心地带。属暖温带大陆性季风气候，四季分明。年平均气温4.8℃，最热月（7月）平均温度高达18.3℃，最冷月（1月）平均温度为-10.1℃。全年无霜期约为195天。年日照2600 h。年降水量为588 mm，6～8月的降水量约占全年降水量的74%，7月多暴雨。样地为20 hm²，东西400 m，南北500 m。地形复杂，最高海拔为1509.3 m，最低海拔为1290 m，平均海拔为1395 m，最大高差为219.3 m（图2-1）。样地地势较陡，坡度范围20°～80°。样地内土壤以山地棕壤为主，pH值在5.71～7.79之间，平均值为6.6，呈微酸性（图2-2）。植被为典型的暖温带落叶阔叶林，群落发育良好。样地内灌木较多，乔木可高达20 m，垂直结构复杂，成层现象较明显，一般可分4层，乔木层、亚乔木层、灌木层和草本层。此外还有藤本植物，交织攀附于乔木和灌木上。

2.1 Basic Information of the Plot

The Donglingshan warm temperate deciduous broad-leaved forest dynamics plot (abbreviated as Donglingshan FDP; 39°95′N, 115°42′E) is located in the center of Xiaolongmen Forest Park Reserve in Qingshui town and close to Beijing Forest Ecosystem Research Station. It belongs to the warm temperate continental monsoon climate with four distinct seasons. The average annual temperature is 4.8℃. The hottest month is July with an average temperature of 18.3℃, and the coldest month is January with an average temperature of -10.1℃. The annual frost-free period is about 195 days. The annual sunshine is 2600 h. The annual precipitation is 588 mm, and the precipitation from June to August accounts for about 74% of the annual precipitation, and in July there is mostly rainstorm. The size of plot is 20 hm², 400 m wide from east to west and 500 m long from north to south. The terrain is complex, the highest elevation is 1509.3 m, the lowest elevation is 1290 m, the average elevation is 1395 m, and the maximum difference of elevation is 219.3 m (Fig. 2-1). The terrain of the plot is relatively steep with a slope ranging from 20° to 80°. The soil in the plot is mainly mountain brown soil and the pH values range from 5.71 to 7.79, with a mean of 6.6 showing slight acidity (Fig. 2-2). The vegetation is a typical warm temperate deciduous broad-leaved secondary forest, and the community grows well. There are many shrubs in the plot, and the trees can reach to 20 m. The vertical structure is complex and the layering phenomenon is obvious. Generally, it can be divided into 4 layers, the canopy layer, the understory layer, the shrub layer and the herb layer. There are also vines, intertwined and clinging to trees and shrubs.

2.2 样地建设与群落调查

2010年5月开始样地建设，2010年9月完成第一次群落调查，随后每5年复查一次。2020年因为新冠疫情，复查工作推迟到2021年。因此截至2021年已经完成3次调查。样地建设和群落调查参照CTFS（Center for Tropical Forest Science）的技术规范，用全站仪把样地分成500个20 m × 20 m的样方，每个20 m × 20 m的样方又分为16个5 m × 5 m的小样方（图2-3）。标定并调查样方内所有胸径≥1 cm的木本植物个体，内容包括植物个体的物种名称、胸径、坐标位置等，准确测量后挂牌标记。

在东灵山暖温带落叶阔叶林20 hm²样地内，均匀设置了150个种子雨收集器（图2-4），收集掉落于种子雨收集器中的种子、凋落物等。并在每个收集器的3个方向（东、北、南），距其2 m处，分别设立1个1 m × 1 m样方，共设立450个（图2-5）。每年于主要生长季7～9月对样方内的草本植物及木本幼苗进行了系统调查，其中草本调查包括记录其名称、高度、盖度、株（丛）数。

2.2 Plot Establishment and Community Investigation

The plot is established and completed the first community investigation in the summer of 2010. Then, each individual of the forest community will be investigated again every 5 years. Because of the breakout of COVID-19 in 2020, we finished the 3rd census until 2021. Following the standard field protocol of CTFS (http://www.ctfs.si.edu), the plot was divided into 500 20 m × 20 m subplots, and each 20 m × 20 m subplot was divided into sixteen 5 m × 5 m quadrates (Fig. 2-3). All freestanding trees in these 5 m × 5 m quadrates with at least 1cm in diameter at breast height (DBH, 1.3 m above the ground) were identified, tagged, and mapped and is accurately measured and tagged.

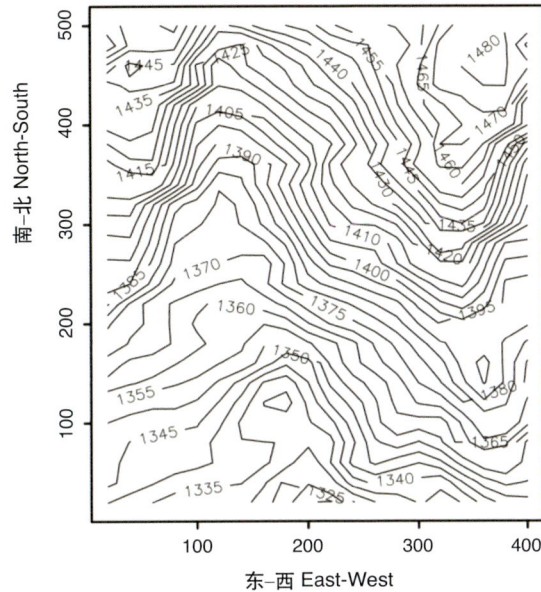

图2-1 东灵山样地等高线图
Fig. 2-1 The topography of the Donglingshan FDP

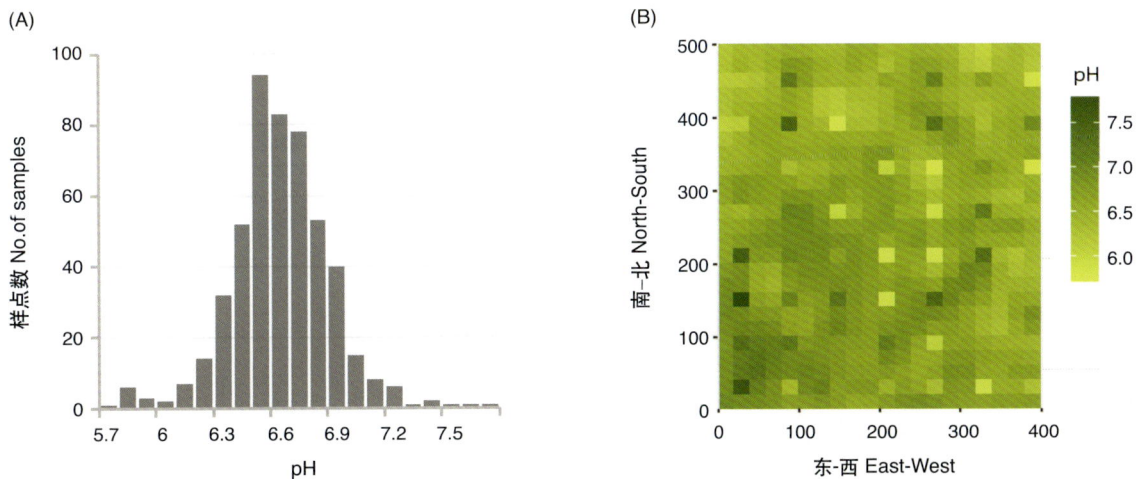

图2-2 土壤pH值直方图（A）及空间分布图（B）
Fig. 2-2 The frequency (A) and spatial distribution (B) of soil pH

There are 150 seed traps evenly set up in the 20 hm² plot to collect seeds, leave litters and others that fell into the collectors (Fig. 2-4). In three directions (east, north, south) of each seed trap, 2 m away from it, we set up three 1 m × 1 m seedling plots respectively, resulting in a total of 450 quadrats (Fig. 2-5). We completely survey the herbaceous plants and woody seedlings in the plots every year from July to September in the main growing season, and record the species name, height, coverage, and number of herbs in each small plot.

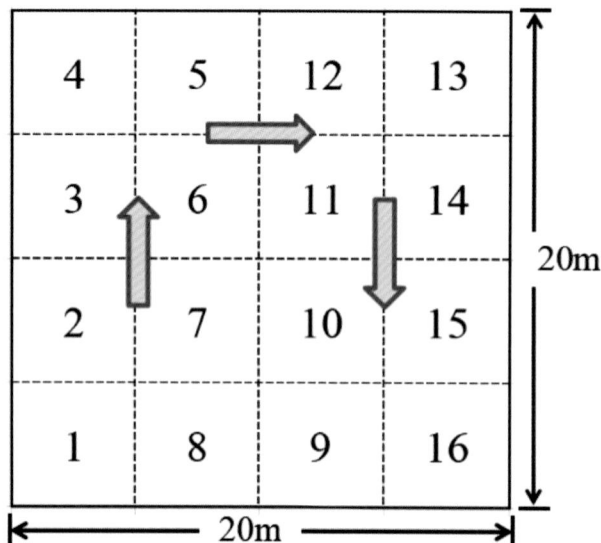

图2-3 20 m × 20 m 样方内5 m × 5 m 小样方设置
Fig. 2-3 The 20 m × 20 m quadrates and 5 m × 5 m subplots surveying woody species in the plot

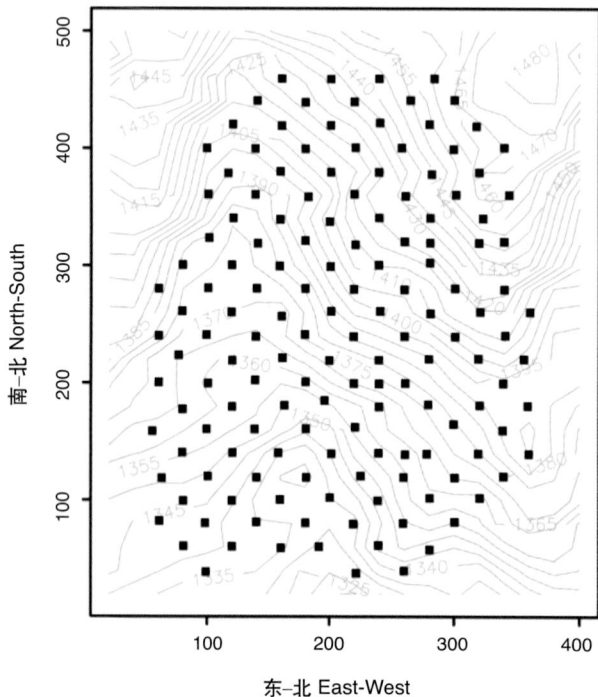

东–北 East-West

图2-4 种子雨收集器分布图（一个黑色方框代表一个收集器，共150个）
Fig. 2-4 The location of 150 seed traps (one black square stands for one seed trap)

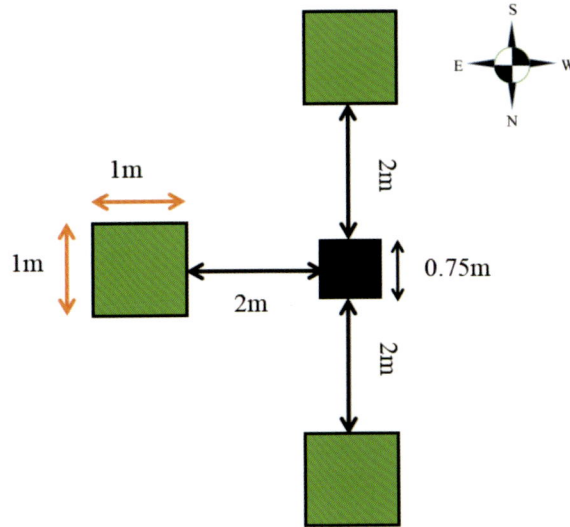

图2-5 幼苗调查样方设置（黑色方框代表收集器，绿色方框代表调查样方）
Fig. 2-5 The seedling plots (black square is seed trap, green squares stands for small plots)

2.3 物种组成与群落结构

根据2021年森林样地复查情况，样地共有47146个胸径（DBH）≥1 cm的独立木本植物个体（包括分枝的个体数为104954），分属于23科33属49种。其中裸子植物1科1属1种，被子植物22科32属48种，涵盖了该地区落叶阔叶林林冠层中大部分常见种，是典型的暖温带落叶阔叶林。按生长型来划分，这些植物中有乔木27种、灌木22种。

2.3 Species Composition and Community Structure

Based on the data analysis of 2021, there were a total of 47146 woody stems with ≥ 1 cm dbh (in total, 104954 records including branches), belonging to 23 families, 33 genera and 49 species. Among them, gymnosperms only had 1 family, 1 genus and 1 species. The other 48 species within 32 genera and 22 families of angiosperms, cover the majority of the common species in the canopy layer of deciduous broad-leaved forests in this region, and are typical species in warm temperate deciduous broad-leaved forests. According to the growth forms, there were 27 tree species and 22 shrub species.

从个体数量上看，有11个物种的个体数超过1000株，其中五角枫（*Acer pictum*）的个体数为11444，其余10个物种按个体数从高到低排序依此是毛榛（*Corylus mandshurica*）、辽东栎、巧玲花（*Syringa pubescens*）、六道木（*Abelia biflora*）、黑桦（*Betula dahurica*）、花曲柳（*Fraxinus chinensis* subsp. *rhynchophylla*）、照山白（*Rhododendron micranthum*）、山杨（*Populus davidiana*）、小花溲疏（*Deutzia parviflora*）、迎红杜鹃（*Rhododendron mucronulatum*）。这11个物种的个体数占样地总个体数的85.82%。此外，有18个物种的个体数为100～1000；有14个物种的个体数为10～100；有6种的个体数小于10。从物种多度累计分布图可以看出（图2-6），个体数最多的前3个物种的个体数超过总个体数的40%，前20个物种的个体数占总个体数的95%，其余29个物种的个体数不到样地总个体数的5%。

In terms of the number of individuals, the number of individuals of 11 species exceeds 1000. Among them, *Acer pictum* had 11444 individuals, and the other 10 species are sorted from the highest to the lowest number of individuals, followed by *Corylus mandshurica, Quercus wutaishanica, Syringa pubescens, Abelia biflora, Betula dahurica, Fraxinus chinensis* subsp. *rhynchophylla, Rhododendron micranthum, Populus davidiana, Deutzia parviflora, Rhododendron mucronulatum*. The number of individuals of these 11 species accounted for 85.82 % of the total individuals in the plot. In addition, the individual number of 18 species was within the range of 100 to 1000; there were 14 species with a number of individuals of 10 to 100; and the individual number of 6 species was less than 10. The distribution of species abundance can be shown in the cumulative distribution map (Fig. 2-6). The number of individuals of 3 species with the top largest number of individuals exceeds 40 % of the total number, the number of individuals of the top 20 species accounts for 95 % of the total number, and the number of individuals in the remaining 29 species was less than 5 % of the total number in the plot.

根据2022年最新的草本植物调查，林下共有草本植物87种，隶属于33科61属，其中菊科植物最多，为19种。其次毛茛科9种，堇菜科6种。3科之和占调查草本总物种数的39%。其中草本的主要优势种有野青茅（*Deyeuxia pyramidalis*）、蒙古风毛菊（*Saussurea mongolica*）和三脉紫菀（*Aster ageratoides*），重要值分别为10.04、9.89、7.47。

According to the latest herbal survey in 2022, there were 87 herbaceous species under the forest, belonging to 33 families and 61 genera. Of which the largest number of species was Asteraceae with 19 species. The second was

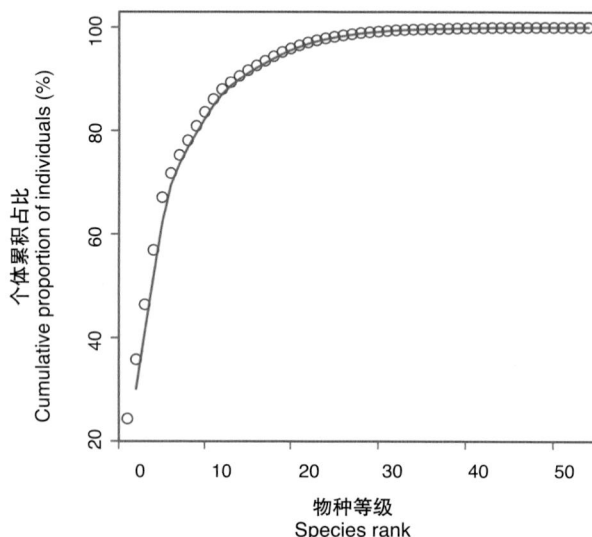

图2-6 物种多度累计分布图
Fig. 2-6 The cumulative distribution curve of species abundance

Ranunculaceae with 9 species, and the third one was Violaceae with 6 species. The sum of the 3 families accounted for 39 % of the total species of herbs investigated. Among them, the main dominant species of herbs were *Deyeuxia pyramidalis*, *Saussurea mongolica* and *Aster trinervius* subsp. *Ageratoides*. The important values of three species was 10.04、9.89 and 7.47 respectively.

样地植被垂直结构清晰，木本植物可分为乔木层、乔木亚层、灌木层。乔木层以五角枫、辽东栎、山杨、白桦等为优势种；乔木亚层以花曲柳等为优势种；灌木层以六道木、巧玲花、迎红杜鹃、照山白、毛榛等为优势种，华北落叶松（*Larix gmelinii* var. *principis-rupprechtii*）是样地唯一一种针叶树种。其中，北京丁香（*Syringa reticulata* subsp. *pekinensis*）和巧玲花是本地区植被温带分布的代表物种（图2-8）。

The vertical structure of the vegetation in the plot is clear, and the woody plants can be divided into tree layer, tree sublayer, and shrub layer. The tree layer is named after the *Acer pictum*, *Quercus wutaishanica*, *Populus davidiana*, *Betula platyphylla* and other dominant species; The small tree layer is dominated by *Fraxinus chinensis* subsp. *rhynchophylla*, etc.; The shrub layer consists of *Abelia biflora*, *Syringa pubescens*, *Rhododendron mucronulatum*, *Rhododendron micranthum*, *Corylus mandshurica* and other dominant species. *Larix gmelinii* var. *principis-rupprechtii* is the only conifer species in the sample. Among them, *Syringa reticulata* subsp. *pekinensis* and *Syringa pubescens* are representative species of temperate vegetation distribution in the region (Fig. 2-8).

样地内木本植物个体胸径最大为63 cm，平均胸径为5.31 cm。样地木本植物个体径级总体呈现倒"J"形分布，胸径介于1～5 cm的个体数占总个体数的67.2 %。随着径级增加，立木株数减少，群落总体上呈现较为良好的更新趋势（图2-7）。

The maximum dbh of woody plants in the plot was 63 cm, and the average dbh was 5.31 cm. The dbh of all woody individual in the plot generally showed an inverted "J" shape distribution, and the number of individuals with dbh no more than 5 cm occupied 67.2 % of the total number of individuals. With the increase of dbh, the number of standing trees decreased, which showed the community regenerate well (Fig. 2-7).

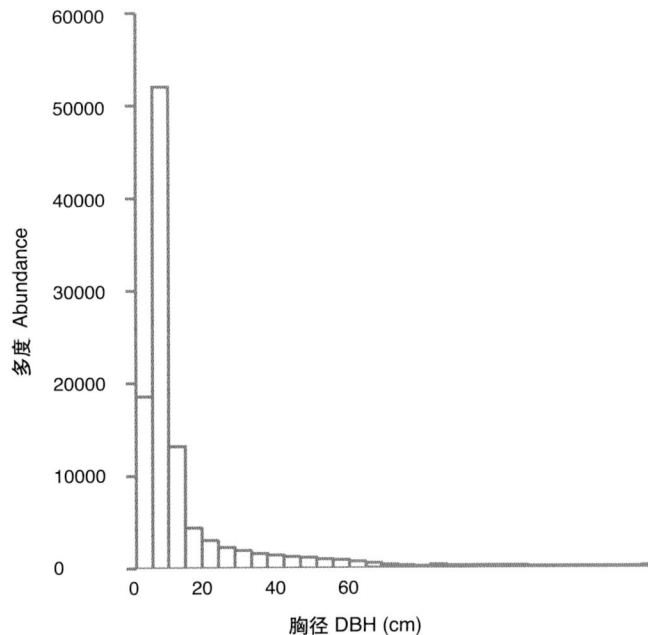

图2-7 木本植物径级结构分布图
Fig. 2-7 The size-class distribution of woody species

图2-8 东灵山样地森林群落和局部影像图片（最后一张无人机图片为刘见礼提供，其余为祝燕提供）
Fig. 2-8 Forest community and partial imagery map of Donglingshan FDP (The last picture taken by drones provided by Jianli Liu and others provided by Yan Zhu)

东灵山样地：木本植物及其分布
Donglingshan Plot:
Woody Plants and Their Distribution

3

1 华北落叶松

huá běi luò yè sōng | Daurian Larch

Larix gmelinii var. *principis-rupprechtii* (Mayr) Pilg.
松科 | Pinaceae

代码 (SpCode) = LARGME

个体数 (Individual number/20 hm²) = 118

最大胸径 (Max DBH) = 20.6 cm

重要值排序 (Importance value rank) = 32

落叶乔木，高达30 m。树冠卵状圆锥形。树皮暗灰褐色，不规则纵裂有鳞。叶条形，长2～3 cm，宽约1 mm，有长短枝之分，长枝上的叶螺旋状互生，短枝上的叶簇生。雌球花生于短枝顶端。球果卵圆体形或圆柱状卵形，种鳞近五角状卵形，边缘不反卷；种翅上部三角状。传粉期4～5月，种子成熟期10月。

Deciduous trees, 30 m tall. Crown ovoid-conical. Bark gray, longitudinally fissured, scaly. Leaves striped, 2-3 cm long, ca. 1 mm wide, there are long and short branches, spirally arranged and sparse on long shoots, in dense clusters on short shoots. Solitary female cones borne at apices of short shoots. Cones ovoid or cylindrically ovate. Seed scales nearly pentagonal-ovate, margin unscrewed; upper wing triangular. Pollination Apr.-May, seed maturity Oct..

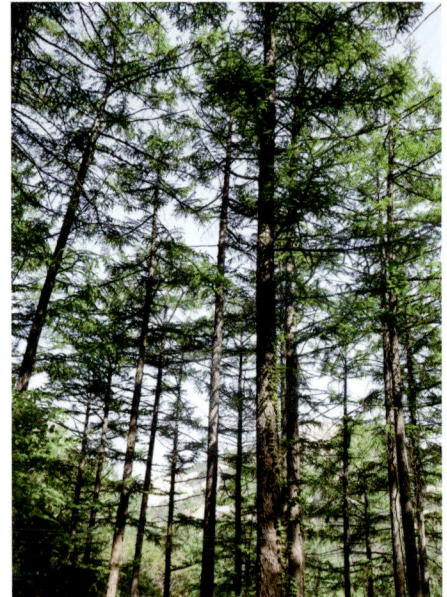

植株　　　　Whole plant
摄影：林秦文　　Photo by: Lin Qinwen

花序　　　　Inflorescences
摄影：林秦文　　Photo by: Lin Qinwen

球果　　　　Cones
摄影：林秦文　　Photo by: Lin Qinwen

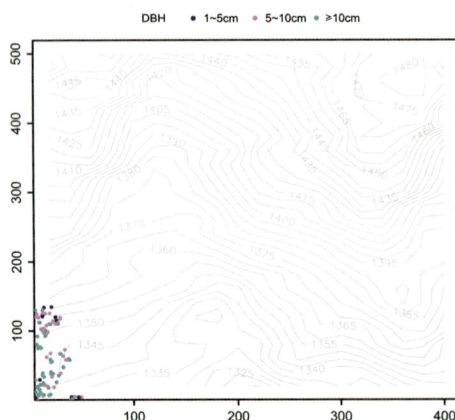

个体分布图 Distribution of individuals

径级分布表 DBH class

胸径区间 (Diameter class) (cm)	个体数 (No. of individuals in the plot)	比例 (Proportion) (%)
1～2	2	1.69
2～5	15	12.71
5～10	55	46.61
10～20	45	38.14
20～30	1	0.85
30～60	0	0.00
≥60	0	0.00

2 东北茶藨子

Ribes mandshuricum (Maxim.) Kom.
茶藨子科 | Grossulariaceae

代码 (SpCode) = RIBMAN
个体数 (Individual number/20 hm^2) = 47
最大胸径 (Max DBH) = 4 cm
重要值排序 (Importance value rank) = 30

落叶灌木，1~3 m。树冠宽卵形。树皮暗褐色。小枝灰褐色，无刺。冬芽卵圆形。叶互生，宽大，掌状3裂，裂片卵状三角形，边缘具粗锯齿，幼时两面被灰白色平贴短柔毛。花两性，总状花序。果红色，无毛。花期4~6月，果期7~8月。

Deciduous shrubs, 1-3 m tall. Crown broadly ovate. Bark gray. Branchlets greyish brown, unarmed. Buds ovoid. The leaves are alternate, broad, palmate with 3 lobes, lobes ovate triangle, with coarsely serrated margins. Leaf appressed-pubescent on both surfaces or adaxially glabrous when young. Flowers bisexual and racemes erect. Fruit globose, red, glabrous. Fl. Apr.-Jun., fr. Jul.-Aug..

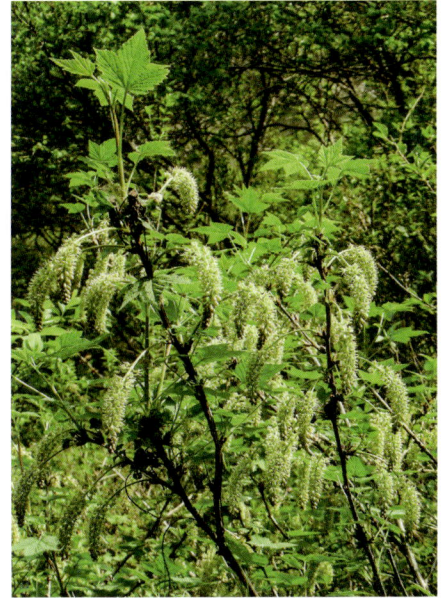

植株　　Whole plant
摄影：林秦文　Photo by: Lin Qinwen

花序　　Inflorescences
摄影：林秦文　Photo by: Lin Qinwen

果序　　Infructescences
摄影：李晓东　Photo by: Li Xiaodong

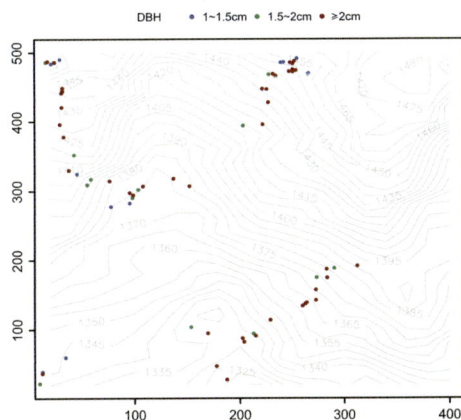

个体分布图 Distribution of individuals

径级分布表 DBH class

胸径区间 (Diameter class) (cm)	个体数 (No. of individuals in the plot)	比例 (Proportion) (%)
1~2	24	51.06
2~5	23	48.94
5~10	0	0.00
10~20	0	0.00
20~30	0	0.00
30~60	0	0.00
≥60	0	0.00

3 兴安胡枝子（达乌里胡枝子）

Lespedeza daurica (Laxm.) Schindl.
豆科 I Fabaceae

代码 (SpCode) = LESDAV

个体数 (Individual number/20 hm^2) = 13

最大胸径 (Max DBH) = 3.9 cm

重要值排序 (Importance value rank) = 41

亚灌木。茎直立，小枝短柔毛。小叶长圆形或狭长圆形，背面贴伏的或直立短柔毛，正面疏生短柔毛或无毛，基部圆形，先端圆形或微缺，短尖。总状花序短于叶，花序梗密被短柔毛；花萼3～6 mm，5深裂，裂片披针形；花冠白色或淡黄白色。荚果倒卵形，短柔毛，先端具喙，包围在宿存花萼内。花期7～8月，果期9～10月。

Subshrubs. Stems often ascending and branchlets pubescent. Leaflets oblong or narrowly oblong, abaxiallyadpressed or erect pubescent, adaxially sparsely pubescent or glabrous, base rounded, apex rounded or emarginate, mucronate. Racemes erect shorter than leaves, peduncle densely pubescent. Calyx 3-6 mm, 5-parted, lobes lanceolate. Corolla white or yellowish white. Lenticular obovoid, pubescent, apex rostrate, enclosed in per-sistent calyx. Fl. Jul.-Aug., fr. Sep.-Oct..

植株　　　　Whole plant
摄影：林秦文　　　Photo by: Lin Qinwen

叶背　　　　Leaves abaxially
摄影：林秦文　　　Photo by: Lin Qinwen

果　　　　Fruits
摄影：林秦文　　　Photo by: Lin Qinwen

径级分布表 DBH class

胸径区间 (Diameter class) (cm)	个体数 (No. of individuals in the plot)	比例 (Proportion) (%)
1～2	8	61.54
2～5	5	38.46
5～10	0	0.00
10～20	0	0.00
20～30	0	0.00
30～60	0	0.00
≥60	0	0.00

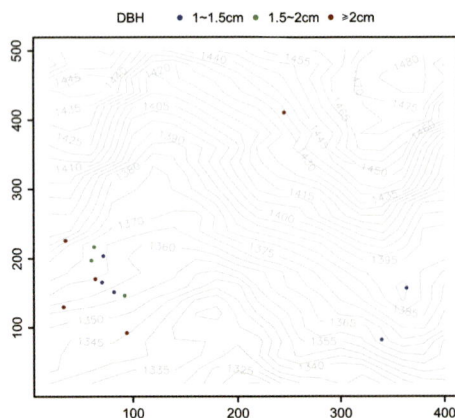

DBH　• 1~1.5cm　• 1.5~2cm　• ≥2cm

个体分布图 Distribution of individuals

4 胡枝子

Lespedeza bicolor Turcz.
豆科 | Leguminosae

代码 (SpCode) = LESBIC
个体数 (Individual number/20 hm^2) = 17
最大胸径 (Max DBH) = 3.2 cm
重要值排序 (Importance value rank) = 38

落叶直立灌木，1～3 m。树冠宽卵形。树皮褐色，皮孔显著。小枝具条棱，被疏短毛。冬芽卵形。羽状复叶互生，质薄，卵形，全缘，下面被疏柔毛。总状花序腋生，比叶长；花萼钟形，5浅裂；花冠蝶形，红紫色，长约1 cm。雄蕊二体，子房被毛。荚果斜倒卵形，稍扁，表面具网纹，密被短柔毛，种子1颗。花期7～9月，果期9～10月。

Deciduous upright shrubs, 1-3 m tall. Crown broadly ovate. Bark gray and lenticels conspicuous. Twigs are ribbed and sparsely haired. Winter buds are ovate. Pinnate compound leaves are alternate, thin, ovate, full margin and underneath the sparse hairs. The inflorescence is axillary and longer than the leaves; calyx bell-shaped, 5 shallow lobes; corolla is butterfly-shaped, reddish-purple and about 1 cm long. Stamens bisomy, ovary indumentum. Pods are obliquely inverted ovate, slightly flattened, with a reticulated surface, densely covered with short soft hairs, and 1 seed. Fl. Jul.-Sep., fr. Sep.-Oct..

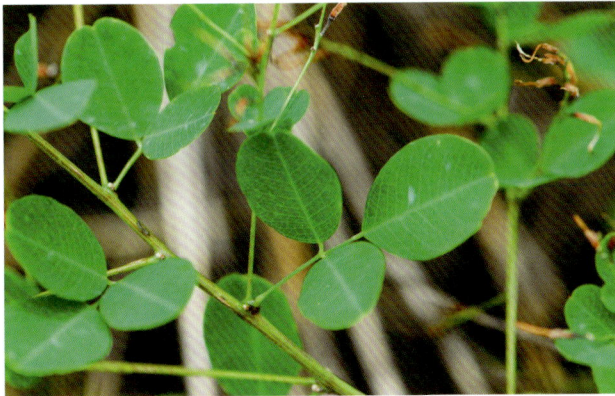

叶 Leaves
摄影：张金政　Photo by: Zhang Jinzheng

花序 Inflorescences
摄影：刘冰　Photo by: Liu Bing

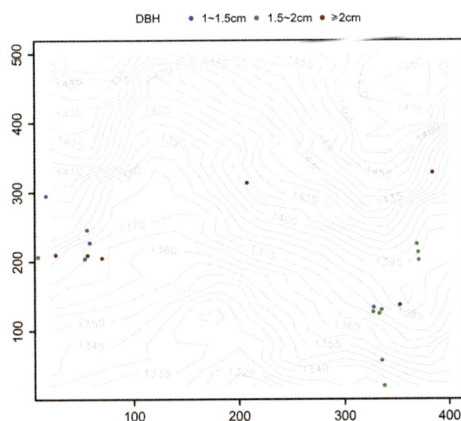

个体分布图 Distribution of individuals

径级分布表 DBH class

胸径区间 (Diameter class) (cm)	个体数 (No. of individuals in the plot)	比例 (Proportion) (%)
1～2	12	70.59
2～5	5	29.41
5～10	0	0.00
10～20	0	0.00
20～30	0	0.00
30～60	0	0.00
≥60	0	0.00

5 土庄绣线菊（柔毛绣线菊）

Spiraea pubescens Turcz.
蔷薇科 | Rosaceae

代码 (SpCode) = SPIPUB
个体数 (Individual number/20 hm^2) = 172
最大胸径 (Max DBH) = 5.3 cm
重要值排序 (Importance value rank) = 66

落叶灌木，1～2 m。树冠宽卵形。树皮灰色。小枝褐黄色，幼时具短柔毛。叶互生，菱状卵形至椭圆形，边缘上部具深刻锯齿，有时3裂，上面具稀疏柔毛，下面被短柔毛。伞形花序具总花梗；花瓣5，白色；花盘波状圆环形。菁葖果开张，仅在腹缝微具短柔毛。花期5～6月，果期7～8月。

Deciduous shrubs, 1-2 m tall. Crown is broadly ovate. Bark gray. Twigs are brownish yellow and have short soft hairs when young. Leaves are alternate, diamond-ovate to oval, with deeply serrated upper margins, sometimes 3 lobes, sparse soft hairs on the upper mask, and short soft hairs on the bottom. The umbel-shaped inflorescence has a total peduncle; petal 5, white; disc is wavy and torused. Fruit is open and has only short soft hairs in the abdominal slit. Fl. May-Jun., fr. Jul.-Aug..

植株　　Whole plant
摄影：林秦文　　Photo by: Lin Qinwen

叶背　　Leaves abaxially
摄影：刘冰　　Photo by: Liu Bing

花序　　Inflorescences
摄影：刘冰　　Photo by: Liu Bing

径级分布表 DBH class

胸径区间 (Diameter class) (cm)	个体数 (No. of individuals in the plot)	比例 (Proportion) (%)
1～2	133	77.33
2～5	38	22.09
5～10	1	0.58
10～20	0	0.00
20～30	0	0.00
30～60	0	0.00
≥60	0	0.00

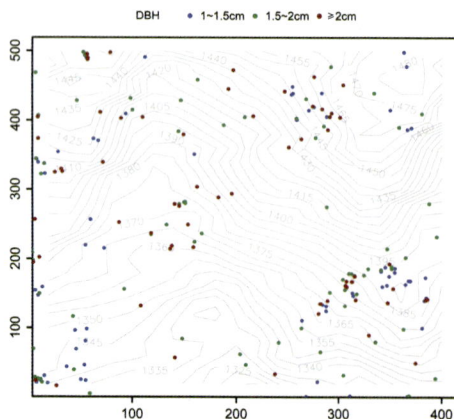

个体分布图 Distribution of individuals

6 灰栒子

huī xún zi | Peking Cotoneasteer

Cotoneaster acutifolius Turcz.
蔷薇科 | Rosaceae

代码 (SpCode) = COTACU
个体数 (Individual number/20 hm²) = 2
最大胸径 (Max DBH) = 2.4 cm
重要值排序 (Importance value rank) = 56

落叶灌木，2~4 m。树冠宽卵形。树皮红褐色，平滑。叶互生，圆卵形，先端急尖，基部宽楔形，全缘。花2~5朵成聚伞花序，被长柔毛；花瓣5，白色外带红晕。梨果椭圆形，黑色。花期5~6月，果期9~10月。

Deciduous shrubs, 2-4 m tall. Crown is broadly ovate, bark is reddish brown and smooth. Leaves are alternate, roundly ovate, sharply pointed at the apex, broadly wedge-shaped at the base, fully margined. 2-5 flowers inflorescences, covered with long soft hairs; petals 5, white with a red halo. The pear fruit is oval in shape, black. Fl. May-Jun., fr. Sep.-Oct..

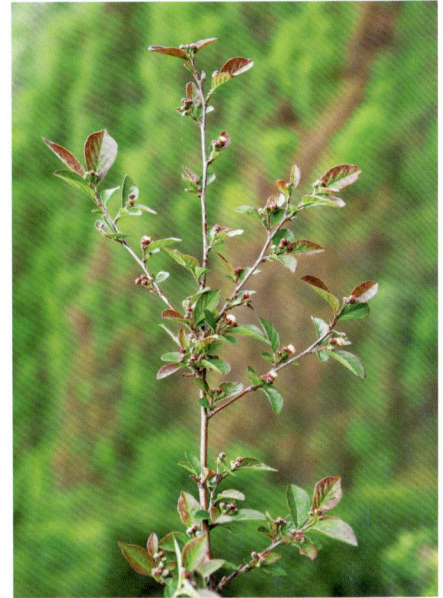

植株　　　　Whole plant
摄影：林秦文　　Photo by: Lin Qinwen

叶背　　　　Leaves abaxially
摄影：张钢民　　Photo by: Zhang Gangmin

果序　　　　Infructescences
摄影：林秦文　　Photo by: Lin Qinwen

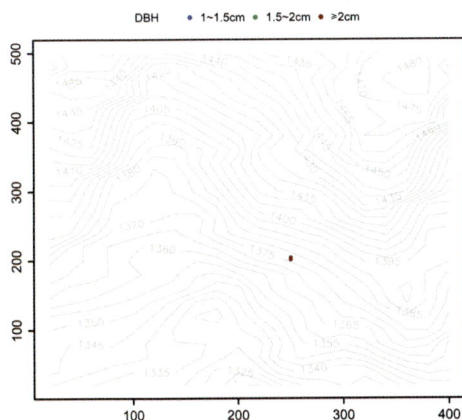

个体分布图 Distribution of individuals

径级分布表 DBH class

胸径区间 (Diameter class) (cm)	个体数 (No. of individuals in the plot)	比例 (Proportion) (%)
1~2	0	0.00
2~5	2	100.00
5~10	0	0.00
10~20	0	0.00
20~30	0	0.00
30~60	0	0.00
≥60	0	0.00

7 北京花楸

Sorbus discolor (Maxim.) Maxim.
蔷薇科 | Rosaceae

代码 (SpCode) = SORDIS

个体数 (Individual number/20 hm^2) = 870

最大胸径 (Max DBH) = 30.9 cm

重要值排序 (Importance value rank) = 12

落叶乔木，达10 m。树冠宽卵形。树皮灰色，平滑。小枝圆柱形，具稀疏皮孔。奇数羽状复叶互生，小叶5～7对，长圆形，边缘有细锐锯齿，无毛。复伞房花序较疏松；萼片5，三角形。花瓣5，卵形，白色；雄蕊15～20。果卵形，白色或黄色，先端具宿存闭合萼片。花期5月，果期8～9月。

Deciduous trees, up to 10 m tall. Crown is broadly ovate. Bark is gray and smooth. Twigs are cylindrical in shape with sparse skin holes. Odd pinnate compound leaves are alternate, with 5-7 pairs of leaflets, oblong, finely sharply serrated margins, and glabrous. The inflorescence of the compound umbrella house is relatively loose; sepals 5, triangular. Petals 5, ovate, white; stamens 15-20. The fruit is ovate, white or yellow, with a closed sepal at the apex. Fl. May, fr. Aug.-Sep..

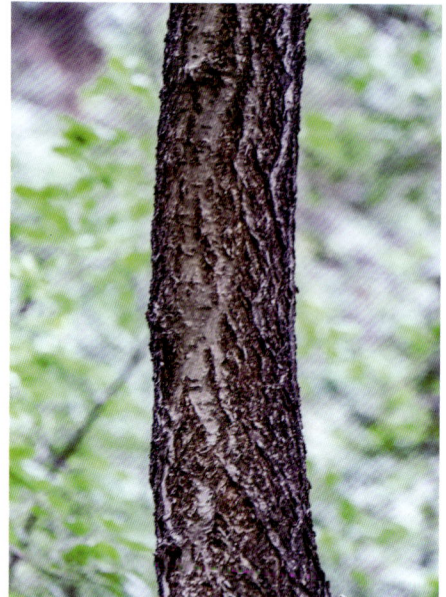

树干　　　　Trunk
摄影：林秦文　Photo by: Lin Qinwen

叶背　　　　Leaves abaxially
摄影：林秦文　Photo by: Lin Qinwen

果序　　　　Infructescences
摄影：张志翔　Photo by: Zhang Zhixiang

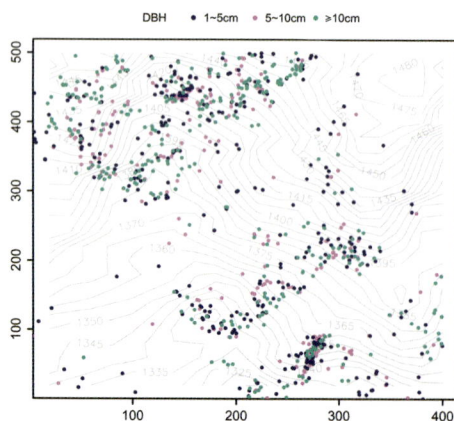

个体分布图 Distribution of individuals

径级分布表　DBH class

胸径区间 (Diameter class) (cm)	个体数 (No. of individuals in the plot)	比例 (Proportion) (%)
1～2	31	3.56
2～5	214	24.60
5～10	312	35.86
10～20	284	32.64
20～30	28	3.22
30～60	1	0.12
≥60	0	0.00

8 山荆子

Malus baccata (L.) Borkh.

蔷薇科 | Rosaceae

代码 (SpCode) = MALBAC

个体数 (Individual number/20 hm^2) = 24

最大胸径 (Max DBH) = 22.5 cm

重要值排序 (Importance value rank) = 36

落叶乔木，达10～14 m。树冠宽卵形。树皮灰白色，浅裂。小枝细弱，微屈曲，红褐色。叶互生，椭圆形或卵形，先端渐尖，边缘具细锐锯齿。伞形花序顶生，具花4～6朵。花瓣5，倒卵形，白色；雄蕊15～20；花柱5或4。梨果近球形，红色或黄色。花期4～6月，果期9～10月。

Deciduous trees, up to 10-14 m tall. Crown is broadly ovate. Bark is off-white, shallowly lobed. Twigs are weak, slightly flexed, reddish-brown. Leaves are alternate, oval or ovate, tapering at the apex, with finely sharp serrated edges. The umbel-shaped inflorescence is apical and has 4-6 flowers. Petals 5, inverted oval, white; stamens 15-20; peduncles 5 or 4. The pear fruit is nearly spherical, red or yellow. Fl. Apr.-Jun., fr. Sep.-Oct..

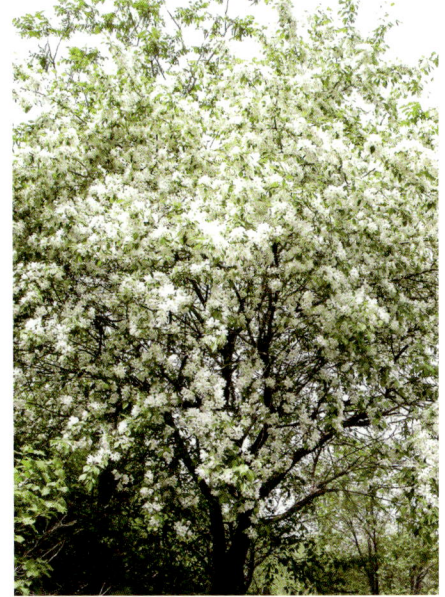

植株　　　　Whole plant
摄影：林秦文　Photo by: Lin Qinwen

花　　　　Flowers
摄影：张志翔　Photo by: Zhang Zhixiang

叶与果序　　Leaves and infructescences
摄影：汪远　　Photo by: Wang Yuan

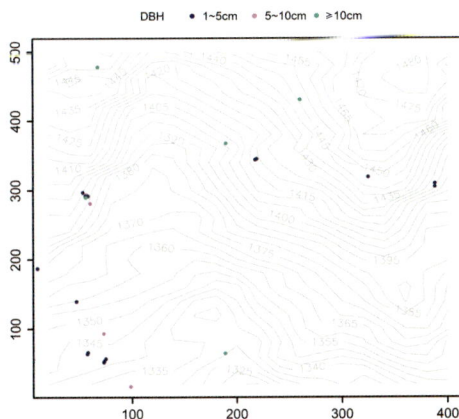

个体分布图 Distribution of individuals

径级分布表 DBH class

胸径区间 (Diameter class) (cm)	个体数 (No. of individuals in the plot)	比例 (Proportion) (%)
1～2	1	4.16
2～5	11	45.83
5～10	7	29.17
10～20	4	16.67
20～30	1	4.17
30～60	0	0.00
≥60	0	0.00

9 山桃

Amygdalus davidiana (Carrière) de Vos ex Henry
蔷薇科 | Rosaceae

代码 (SpCode) = AMYDAV

个体数 (Individual number/20 hm^2) = 247

最大胸径 (Max DBH) = 45.6 cm

重要值排序 (Importance value rank) = 23

落叶乔木，达10 m。树冠宽卵形。树皮平滑，具环纹，常暗紫色。叶互生，卵状披针形，先端长渐尖，基部宽楔形，边缘具细锐锯齿，两面无毛。花单生，先叶开放，近无梗；花瓣粉红色或白色。核果球形，有沟，有毛，离核。花期3～4月，果期7～8月。

Deciduous trees, up to 10 m tall. Crown is broadly ovate. Bark is smooth, ringed, and often dark purple. Leaves are alternate, ovate lanceolate, with a long tapering apex, a broad wedge at the base, finely sharp serrated edges, and glabrous on both sides. Flowers are solitary, open first leaves, nearly installed; the petals are pink or white. Drupe is spherical, grooved, hairy, off-core. Fl. Mar.-Apr., fr. Jul.-Aug..

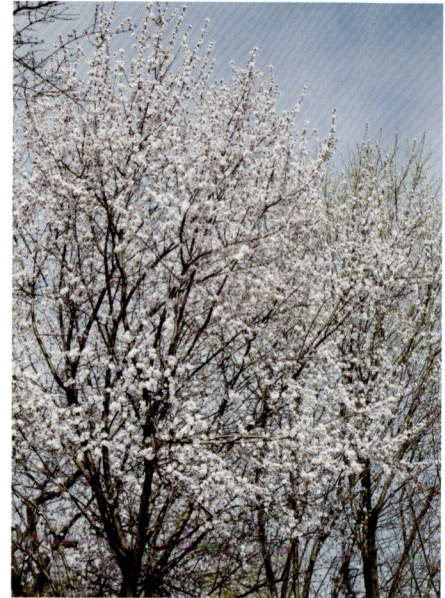

植株　　Whole plant
摄影：叶剑飞　　Photo by: Ye Jianfei

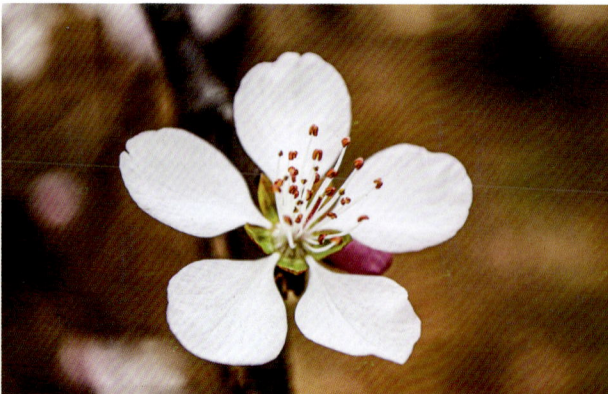

花　　Flower
摄影：叶建飞　　Photo by: Ye Jianfei

叶与果　　Leaves and fruits
摄影：林秦文　　Photo by: Lin Qinwen

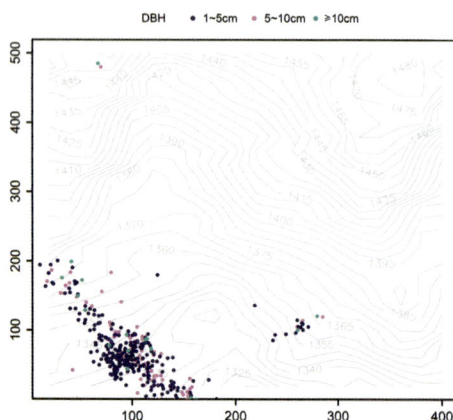

个体分布图 Distribution of individuals

径级分布表 DBH class

胸径区间 (Diameter class) (cm)	个体数 (No. of individuals in the plot)	比例 (Proportion) (%)
1～2	3	1.21
2～5	116	46.97
5～10	119	48.18
10～20	7	2.84
20～30	1	0.40
30～60	1	0.40
≥60	0	0.00

10　山杏

Armeniaca sibirica (L.) Lam.
蔷薇科 | Rosaceae

代码 (SpCode) = ARMSIB
个体数 (Individual number/20 hm²) = 24
最大胸径 (Max DBH) = 13.8 cm
重要值排序 (Importance value rank) = 34

落叶灌木或小乔木，2～5 m。树冠宽卵形。树皮粗糙，纵裂，暗灰色。叶互生，卵形或近圆形，先端长渐尖至尾尖，叶边有细钝锯齿，两面无毛。花单生，5数，先叶开放；花瓣近圆形，白色或粉红色。核果扁球形，熟时黄色或橘红色，被毛。花期3～4月，果期6～7月。

Deciduous shrubs or small trees, 2-5 m tall. Crown broadly ovate. Bark is rough, longitudinally lobed and dark grey. Leaves are alternate, ovate or nearly rounded, with a long tapering apex to the tip of the tail, fine blunt serrations on the leaf edges, and hairless on both sides. Flowers are solitary, number 5, and the first leaves are open; petals nearly round, white or pink. The drupes are flattened and spherical, yellow or orange-red when ripe, and hairy. Fl. Mar.-Apr., fr. Jun.-Jul..

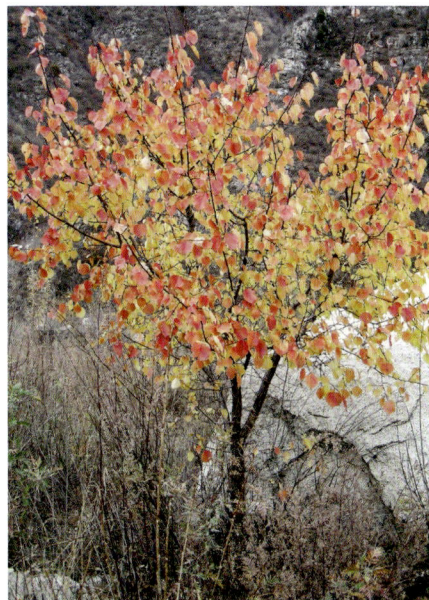

植株　　　Whole plant
摄影：叶剑飞　　　Photo by: Ye Jianfei

花　　　Flowers
摄影：林秦文　　　Photo by: Lin Qinwen

果实　　　Fruits
摄影：林秦文　　　Photo by: Lin Qinwen

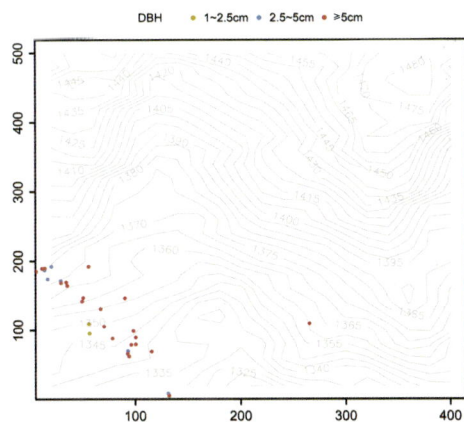

个体分布图 Distribution of individuals

径级分布表　DBH class

胸径区间 (Diameter class) (cm)	个体数 (No. of individuals in the plot)	比例 (Proportion) (%)
1～2	1	4.17
2～5	4	16.67
5～10	11	45.83
10～20	8	33.33
20～30	0	0.00
30～60	0	0.00
≥60	0	0.00

11 稠李 chóu lǐ | Bird Cherry

Padus avium Mill.
蔷薇科 I Rosaceae

代码 (SpCode) = PRUPAD
个体数 (Individual number/20 hm^2) = 52
最大胸径 (Max DBH) = 12.2 cm
重要值排序 (Importance value rank) = 31

落叶乔木，达15 m。树冠宽卵形。树皮暗褐色，皮孔显著；小枝具棱，紫褐色。叶互生，椭圆形至倒卵形，边缘有锐锯齿。叶柄近顶端有2腺体。总状花序下垂；花瓣5，白色，有香味，倒卵形。核果球形或卵球形，黑色，有光泽；核有明显皱纹。花期4~5月，果期9~10月。

Deciduous trees, up to 15 m tall. Crown broadly ovate. Bark is dark brown with prominent skin holes. Twigs are ridged and purple-brown. Leaves are alternate, oval to inverted ovate, with sharp serrations at the edges. There are 2 glands near the apex of the petiole. The inflorescence droops; petals 5, white, scented, inverted ovate. Drupes are spherical or ovoid in shape, black, shiny; nucleus has obvious wrinkles. Fl. Apr.-May, fr. Sep.-Oct..

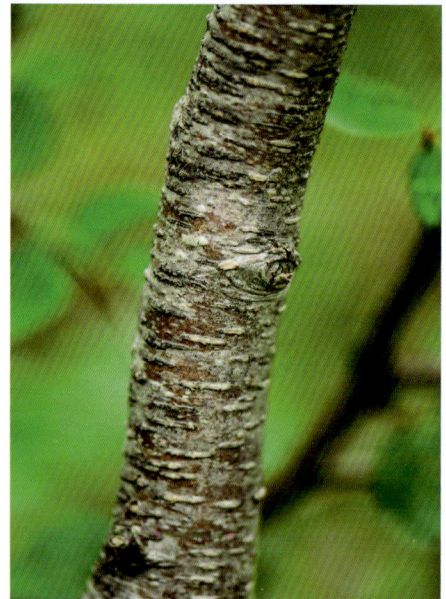

树干　　　　　　Trunk
摄影：张志翔　　Photo by: Zhang Zhixiang

叶与花序　　　　Leaves and inflorescences
摄影：刘冰　　　Photo by: Liu Bing

果序　　　　　　Infructescences
摄影：林秦文　　Photo by: Lin Qinwen

径级分布表 DBH class

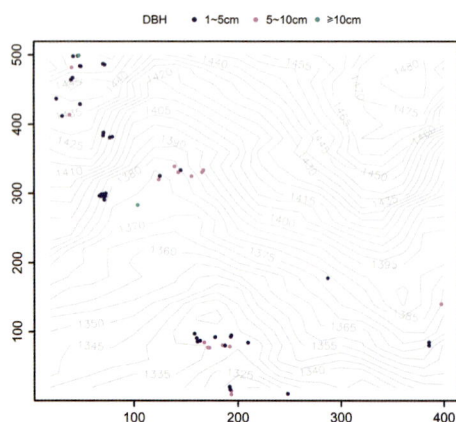

个体分布图 Distribution of individuals

胸径区间 (Diameter class) (cm)	个体数 (No. of individuals in the plot)	比例 (Proportion) (%)
1~2	3	5.77
2~5	25	48.07
5~10	22	42.31
10~20	2	3.85
20~30	0	0.00
30~60	0	0.00
≥60	0	0.00

12 小叶鼠李

Rhamnus parvifolia Bunge
鼠李科 | Rhamnaceae

代码 (SpCode) = RHAPAR

个体数 (Individual number/20 hm^2) = 122

最大胸径 (Max DBH) = 12.3 cm

重要值排序 (Importance value rank) = 27

落叶灌木，1.5～2 m。树冠卵形。树皮灰褐色。小枝紫褐色，具针刺。冬芽卵形。叶对生至簇生，纸质，菱形，顶端钝，基部楔形，边缘具圆齿状细锯齿。叶柄长4～15 mm。花簇生于短枝；花小，单性，雌雄异株，黄绿色。花萼4裂。花瓣4。雄蕊4，为花瓣抱持。花盘薄，杯状。子房上位，球形。核果倒卵状球形，熟时黑色，2核；种子倒卵圆形，背侧有纵沟。花期4～5月，果期6～9月。

Deciduous shrubs, 1.5-2 m tall. Crown ovate. Bark is grey-brown. Twigs purple-brown with pinpricks. Winter buds are ovate. Leaves are paraphyletic to clustered, papery, diamond-shaped, blunt at the apex, wedge-shaped at the base, and finely serrated with rounded teeth at the edges. Petiole is 4-15 mm long. Flower clusters grow on short branches; small, unisexual, hermaphroditic, yellow-green flowers. Calyx 4 clefts. Petal 4. Stamens 4, held for petals. The flower tray is thin, cup-shaped. The ovary is superior and spherical. Drupe is inverted ovate and spherical, black when ripe, with 2 cores; seeds are inverted ovate with longitudinal grooves on the dorsal side. Fl. Apr.-May, fr. Jun.-Sep..

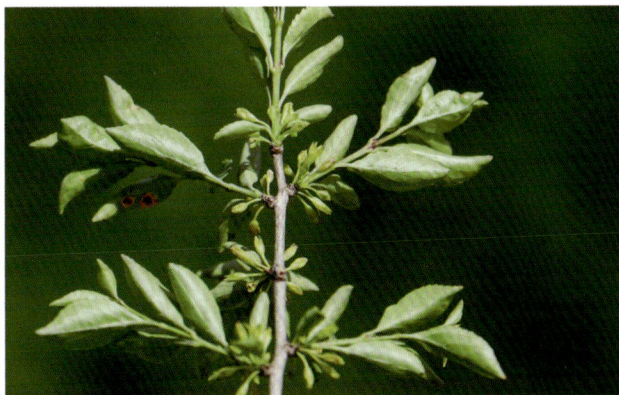

叶与花序　　Leaves and inflorescences
摄影：林秦文　　Photo by: Lin Qinwen

叶与果　　Leaves and fruits
摄影：林秦文　　Photo by: Lin Qinwen

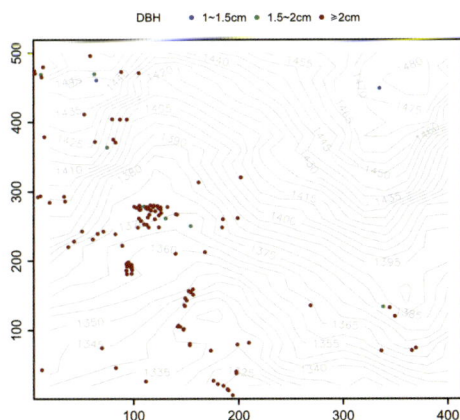

个体分布图　Distribution of individuals

径级分布表　DBH class

胸径区间 (Diameter class) (cm)	个体数 (No. of individuals in the plot)	比例 (Proportion) (%)
1～2	14	11.48
2～5	86	70.49
5～10	19	15.57
10～20	3	2.46
20～30	0	0.00
30～60	0	0.00
≥60	0	0.00

Rhamnus davurica Pall.
鼠李科 | Rhamnaceae

代码 (SpCode) = RHADAV
个体数 (Individual number/20 hm^2) = 566
最大胸径 (Max DBH) = 22 cm
重要值排序 (Importance value rank) = 16

落叶小乔木或灌木，高达10 m。树冠卵形。树皮暗灰褐色。小枝粗壮，无刺；冬芽顶芽大型，卵状披针形。叶对生或束生枝端，卵状椭圆形至倒宽披针形，无毛，边缘具细圆齿。叶柄长1～2 cm。花3～5束生于叶腋；花小，单性，淡绿色。花萼钟状，4裂。雄蕊4，为花瓣抱持。花盘薄，杯状。子房上位，球形。核果球形，熟时黑紫色，直径6 mm，2核；种子卵形，背面有沟。花期5～6月，果期8～9月。

Deciduous small trees or shrubs, up to 10 m tall. Crown ovate. Dark grey-brown bark. Stout twigs, without thorns. The top buds of the winter buds are large and ovate lanceolate. Leaves are opposite or bundled branches, ovate oval to inverted wide lanceolate, glabrous, with fine rounded teeth at the edges. Petiole is 1-2 cm long. 3-5 bunches of flowers grow in leaf axils; flowers are small, unisexual, pale green. Calyx bell-shaped, 4 lobes. Stamens 4, held for petals. The flower tray is thin, cup-shaped. Ovary is superior and spherical. Drupe is spherical, black-purple when ripe, 6 mm in diameter, with 2 cores; seeds are ovate with grooves on the back. Fl. May-Jun., fr. Aug.-Sep..

花序　　Inflorescences
摄影：林秦文　　Photo by: Lin Qinwen

叶与果序　　Leaves and infructescences
摄影：林秦文　　Photo by: Lin Qinwen

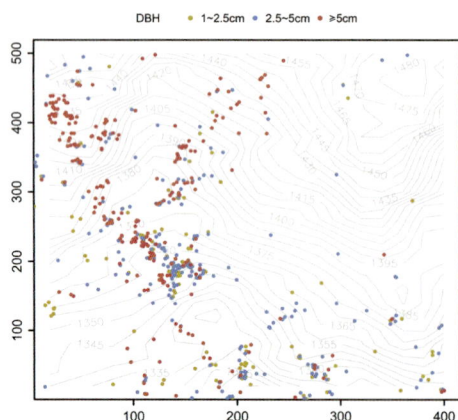

个体分布图 Distribution of individuals

径级分布表　DBH class

胸径区间 (Diameter class) (cm)	个体数 (No. of individuals in the plot)	比例 (Proportion) (%)
1～2	43	7.60
2～5	274	48.41
5～10	181	31.98
10～20	67	11.84
20～30	1	0.17
30～60	0	0.00
≥60	0	0.00

14 东北鼠李

Rhamnus schneideri var. *manshurica* Nakai

鼠李科 | Rhamnaceae

代码 (SpCode) = RHASCH

个体数 (Individual number/20 hm²) = 23

最大胸径 (Max DBH) = 9.5 cm

重要值排序 (Importance value rank) = 46

灌木。小枝无毛，枝端具刺。芽卵圆形，鳞片有缘毛。叶互生，或在短枝上簇生，倒卵形，长2.5～6 cm，先端突尖、短渐尖或渐尖，稀尖，基部楔形或近圆，有圆齿，上面被白色柔毛，下面沿脉或脉腋被疏柔毛。花单性异株，4基数，有花瓣；常数朵至十余朵族生短枝。雌花花梗长0.9～1.3 cm，无毛。萼片披针形，长约3 mm，常反折。花柱2裂。核果球形。花期5～6月，果期7～10月。

Shrubs. Twigs are glabrous and the branches are prickly. Buds are ovoid and the scales are marginal hairs. Leaves are alternate, or clustered on short branches, inverted ovate, 2.5-6 cm long, with a precipitous apex, a short taper or a taper, a thin tip, a wedge-shaped or nearly circular at the base, with rounded teeth, covered with white soft hairs above and sparsely hairy along the veins or axillae veins below. Flowers unisexual heterogeneous, 4 cardinal, with petals; constant flowers to more than 10 flowers grow short branches. The female peduncle is 0.9-1.3 cm long and glabrous. Sepals are lanceolate, about 3 mm long, often inverted. Peduncle is 2 lobes. Drupes are spherical. Fl. May-Jun., fr. Jul.-Oct..

花序　　　　　　　　　Inflorescences
摄影：林秦文　　　　　Photo by: Lin Qinwen

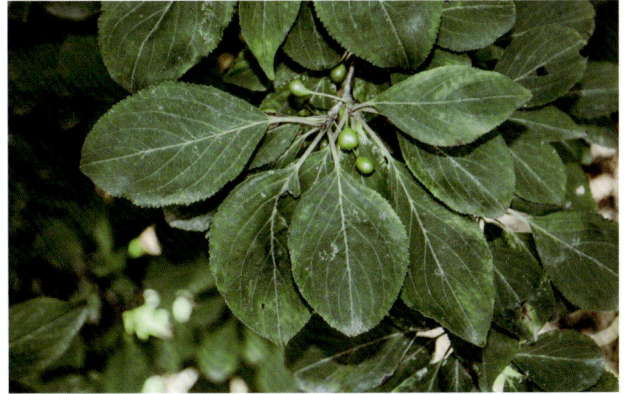

叶与果　　　　　　　　Leaves and fruits
摄影：林秦文　　　　　Photo by: Lin Qinwen

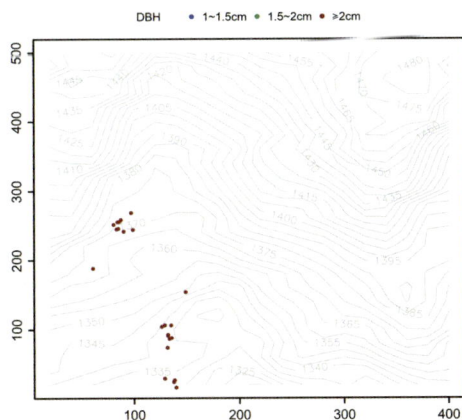

个体分布图 Distribution of individuals

径级分布表　DBH class

胸径区间 (Diameter class) (cm)	个体数 (No. of individuals in the plot)	比例 (Proportion) (%)
1～2	0	0.00
2～5	20	86.96
5～10	3	13.04
10～20	0	0.00
20～30	0	0.00
30～60	0	0.00
≥60	0	0.00

Ulmus davidiana var. *japonica* (Rehder) Nakai
榆科 | Ulmaceae

代码 (SpCode) = ULMDAV

个体数 (Individual number/20 hm²) = 348

最大胸径 (Max DBH) = 81 cm

重要值排序 (Importance value rank) = 18

落叶乔木，3～10 m。树冠宽卵形。树皮暗灰色，具纵裂纹。枝常具木栓质翅。叶互生，倒卵形，边缘具重锯齿，下面脉腋常有毛簇。聚伞花序簇生叶腋，具短梗，先叶开放。花被钟形，4～6浅裂，膜质。翅果长圆状倒卵形，无毛。花期3～4月，果期4～6月。

Deciduous trees, 3-10 m tall. Crown broadly ovate. Bark blackish gray, longitudinally fissured. Branchlets sometimes with corky wings. Leaves alternate, blade obovate, margin doubly serrate. The axils of the veins are often hairy. Inflorescences fascicled cymes on base of branchlets, first leaves open; pedicels short. Perianth campanulate, 4-6 lobed, membrane. Samaras oblong obovate, glabrous. Fl. Mar.-Apr., fr. Apr.-Jun..

树干   Trunk
摄影：林秦文   Photo by: Lin Qinwen

叶与果序   Leaves and infructescences
摄影：林秦文   Photo by: Lin Qinwen

花序   Inflorescences
摄影：林秦文   Photo by: Lin Qinwen

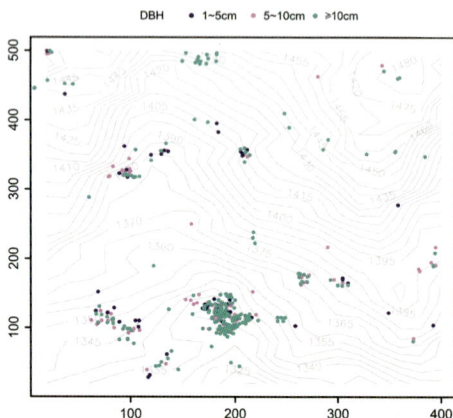

个体分布图 Distribution of individuals

径级分布表 DBH class

胸径区间 (Diameter class) (cm)	个体数 (No. of individuals in the plot)	比例 (Proportion) (%)
1～2	3	0.89
2～5	40	11.50
5～10	114	32.80
10～20	165	47.51
20～30	19	5.56
30～60	6	1.73
≥60	1	0.01

16 大果榆

Ulmus macrocarpa Hance
榆科 | Ulmaceae

代码 (SpCode) = ULMMAC

个体数 (Individual number/20 hm^2) = 417

最大胸径 (Max DBH) = 31.8 cm

重要值排序 (Importance value rank) = 17

落叶乔木或灌木，高达20 m。树冠宽卵形。树皮暗灰色，具纵裂纹。枝常具木栓质翅，小枝淡黄褐色。叶互生，宽倒卵形，革质，边缘具重锯齿。聚伞花序簇生叶腋，先叶开放；花被钟形，膜质。翅果近圆形，两面和边缘被毛，果核位于翅果中部。花期3~4月，果期4~6月。

Deciduous trees or shrubs, up to 20 m tall. Crown broadly ovate. Bark gray to blackish gray, longitudinally fissured. Branchlets light brown, sometimes with corky wings. Leaves alternate, blade broadly obovate, leathery, margin doubly serrate. Inflorescences fascicled cymes on base of branchlets, first leaves open; perianth campanulate, membrane. Samaras orbicular, two sides and edges hairy. Seeds at center of samara. Fl. Mar.-Apr., fr. Apr.-Jun..

树干　　　　　Trunk
摄影：褚建民　　Photo by: Chu Jianmin

叶与翅果　　　Leaves and samaras
摄影：林秦文　　Photo by: Lin Qinwen

木栓质翅　　　Corky wings
摄影：林秦文　　Photo by: Lin Qinwen

径级分布表　DBH class

胸径区间 (Diameter class) (cm)	个体数 (No. of individuals in the plot)	比例 (Proportion) (%)
1~2	9	2.16
2~5	93	22.30
5~10	140	33.57
10~20	146	35.01
20~30	26	6.24
30~60	3	0.72
≥60	0	0.00

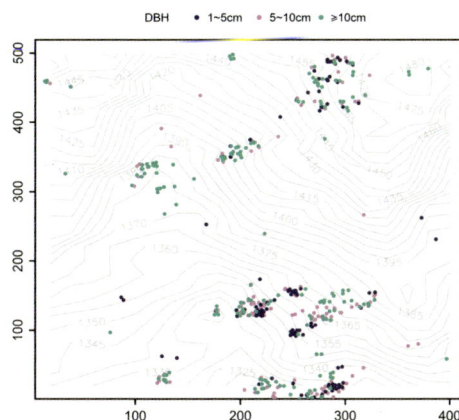

个体分布图 Distribution of individuals

17 裂叶榆

Ulmus laciniata (Herder) Mayr ex Schwapp.
榆科 I Ulmaceae

代码 (SpCode) = ULMLAC

个体数 (Individual number/20 hm^2) = 311

最大胸径 (Max DBH) = 55.2 cm

重要值排序 (Importance value rank) = 20

落叶乔木，高达27 m。树皮淡灰褐色或灰色，浅纵裂。叶互生，倒卵形，通常3~7裂，渐尖或尾状；叶面密生硬毛，粗糙；叶背被柔毛，沿叶脉较密，脉腋常有簇生毛。花小，聚伞花序簇生于上年生枝叶腋。翅果椭圆形。花期4~5月，果期5~6月。

Deciduous trees, up to 27 m tall. Bark dark grayish-brown to gray, longitudinally fissured, exfoliating in flakes. Leaves alternate, obovate, usually 3-7 lobed, base oblique, acuminate or caudate. Leaves densely hard-haired, rough; dorsal pilose, dense along veins, often clustered in axils. Flowers small. Inflorescences fascicled cymes on second year branchlets. Samaras elliptic. Fl. Apr.-May, fr. May-Jun..

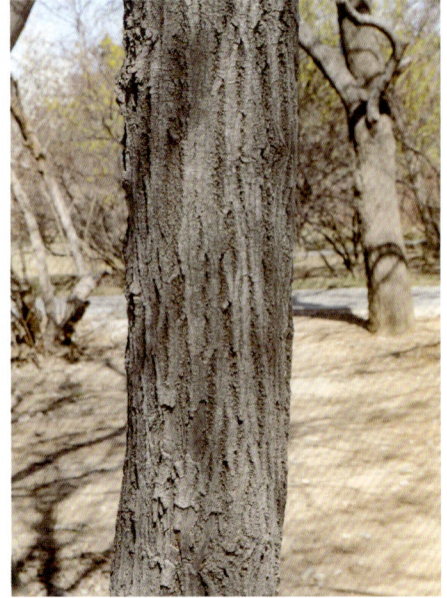

树干　　　　　Trunk
摄影：林秦文　Photo by: Lin Qinwen

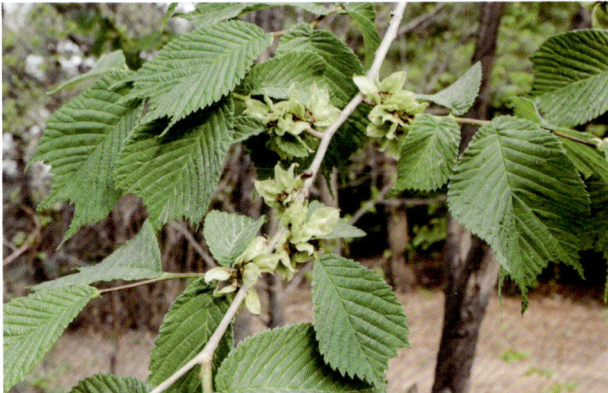

叶与翅果　　　Leaves and samaras
摄影：林秦文　Photo by: Lin Qinwen

花序　　　　　Inflorescences
摄影：林秦文　Photo by: Lin Qinwen

径级分布表　DBH class

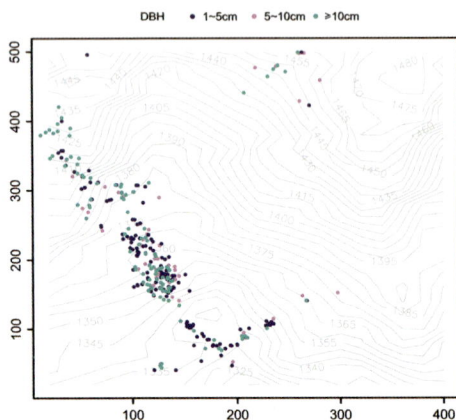

个体分布图 Distribution of individuals

胸径区间 (Diameter class) (cm)	个体数 (No. of individuals in the plot)	比例 (Proportion) (%)
1~2	36	11.57
2~5	105	33.76
5~10	63	20.26
10~20	60	19.29
20~30	36	11.58
30~60	11	3.54
≥60	0	0.00

18 辽东栎

Quercus wutaishanica Mayr
壳斗科 | Fagaceae

代码 (SpCode) = QUEWUT

个体数 (Individual number/20 hm^2) = 4900

最大胸径 (Max DBH) = 79.3 cm

重要值排序 (Importance value rank) = 1

落叶乔木，达30 m。树冠呈卵形。树皮灰褐色，纵裂。小枝无毛，紫褐色。顶芽长卵形。单叶互生，近无柄，倒卵形，侧脉7～12对，直伸，边缘深波状，具圆钝锯齿。花单性，雌雄同株。雄花序为下垂荑花序；雌花单生于总苞内，子房3室。壳斗杯形，包着坚果1/3～1/2，径1.5～1.8 cm。小苞片三角状卵形，呈半球形瘤状突起，密被灰白色短绒毛。坚果卵形。花期5～6月，果期9～10月。

Deciduous trees, up to 30 m tall. Crown ovate. Bark grayish brown, longitudinally fissured. Branchlets purple-brown, glabrous. Terminal bud long ovate. Leaves nearly sessile, obovate, distal widest, lateral veins 7-12 pairs, each terminal at a lobe. Flowers unisexual. Male spike multiflowered, yellowish-green, pendulous; female flowers axillary, solitary in involucrums. Ovary 3-loculed, capitate cup-shaped, enclosing 1/3-1/2 of nut, diameter 1.5-1.8 cm. Bracts triangular-ovate, hemispherical tuberculate, densely grayish-white pubescent. Nuts ovate. Fl. May-Jun., fr. Sep.-Oct..

叶与果　　　　　　　　　Leaves and fruits
摄影：林秦文　　　　　　　Photo by: Lin Qinwen

花序　　　　　　　　　　Inflorescences
摄影：林秦文　　　　　　　Photo by: Lin Qinwen

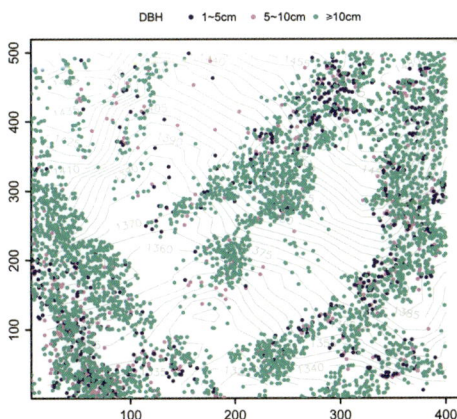

个体分布图 Distribution of individuals

径级分布表 DBH class

胸径区间 (Diameter class) (cm)	个体数 (No. of individuals in the plot)	比例 (Proportion) (%)
1～2	41	0.85
2～5	452	9.23
5～10	665	13.58
10～20	1460	29.80
20～30	1688	34.46
30～60	591	12.07
≥60	3	0.01

19 胡桃楸

Juglans mandshurica Maxim.
胡桃科 | Juglandaceae

代码 (SpCode) = JUGMAN

个体数 (Individual number/20 hm^2) = 505

最大胸径 (Max DBH) = 54.1 cm

重要值排序 (Importance value rank) = 11

落叶乔木，高达30 m。树冠宽卵形。树皮灰色，具浅纵裂。奇数羽状复叶互生，大型。叶长圆形至矩圆状椭圆形，先端尖，边缘具细锯齿。雄柔荑花序腋生下垂，雌穗状花序于新枝顶生。果序俯垂。假核果5~7个，卵球形，顶端尖，密被腺质短柔毛；果核具6~8条纵脊及雕刻状花纹。花期5月，果期8~9月。

Deciduous trees, up to 30 m tall. Crown broadly ovate. Bark gray, shallowly striate. Compound leaves large, odd-pinnate. Leaflets sessile, elliptic to long-elliptic, apex pointed, margin serrate. Male spikes solitary, pendulous. Female spike erect. Fruiting spike pendulous. Nuts 5-7, ovoid, husk densely glandular, pubescent, shell thick, rough, with 6-8 prominent ridges and deep pits and depressions. Fl. May, fr. Aug.-Sep..

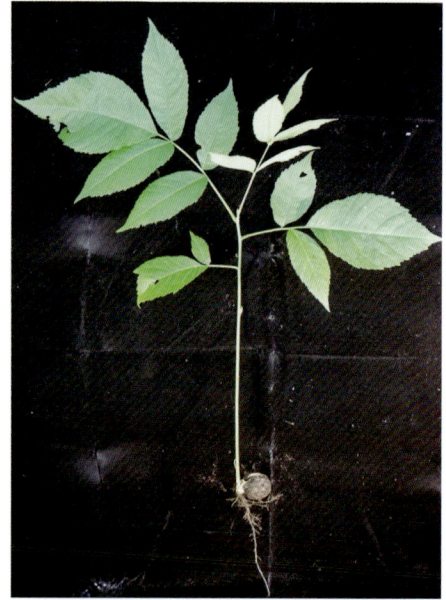

叶　　Leaves
摄影：林秦文　　Photo by: Lin Qinwen

叶与花序　　Leaves and inflorescences
摄影：林秦文　　Photo by: Lin Qinwen

果　　Fruits
摄影：林秦文　　Photo by: Lin Qinwen

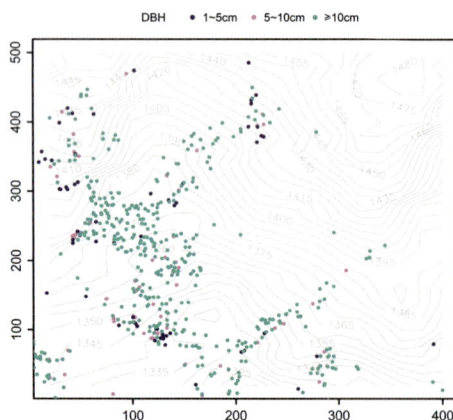

个体分布图 Distribution of individuals

径级分布表 DBH class

胸径区间 (Diameter class) (cm)	个体数 (No. of individuals in the plot)	比例 (Proportion) (%)
1~2	7	1.39
2~5	49	9.70
5~10	58	11.48
10~20	131	25.94
20~30	143	28.32
30~60	117	23.17
≥60	0	0.00

20 白桦

bái huà | Asian White Birch

Betula platyphylla Sukaczev
桦木科 | Betulaceae

代码 (SpCode) = BETPLA

个体数 (Individual number/20 hm²) = 631

最大胸径 (Max DBH) = 57 cm

重要值排序 (Importance value rank) = 8

落叶乔木，10～20 m。树冠宽卵形。树皮灰白色，具白粉，近光滑，通常不剥裂。小枝红褐色，具腺点，无毛。单叶互生，具柄，菱状卵形，先端渐尖，边缘具重锯齿。花单性，雌雄同株，均组成柔荑花序。果序单生，圆柱状，显著下垂；翅果狭椭圆形。花期5～6月，果期8～9月。

Deciduous trees, 10-20 m tall. Crown broadly ovate. Bark grayish white, exfoliating in sheets. Branchlets reddish brown, sparsely resinous, glandular, glabrous. Leaves alternate, ovate-triangular, apex acuminate, margin serraten. Monoecious, male and female flowers in multi-flowered catkins. Infructescence solitary, terete, conspicuously pendulous; nutlet oblong. Fl. May-Jun., fr. Aug.-Sep..

花序　　　　　　　　Inflorescences
摄影：林秦文　　　　Photo by: Lin Qinwen

树干　　　　　　　　Trunk
摄影：汪远　　　　　Photo by: Wang Yuan

翅果　　　　　　　　Samaras
摄影：褚建民　　　　Photo by: Chu Jianmin

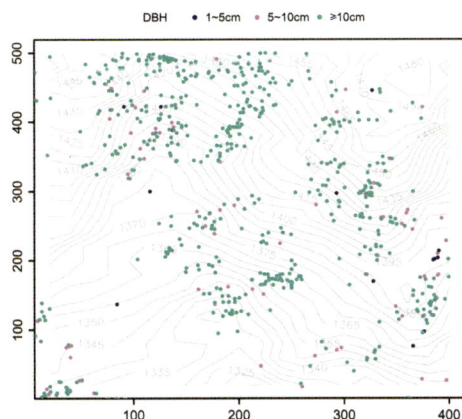

个体分布图 Distribution of individuals

径级分布表　DBH class

胸径区间 (Diameter class) (cm)	个体数 (No. of individuals in the plot)	比例 (Proportion) (%)
1～2	3	0.48
2～5	10	1.58
5～10	88	13.95
10～20	239	37.88
20～30	201	31.85
30～60	90	14.26
≥60	0	0.00

21 黑桦

hēi huà | Dahurian Birch

Betula dahurica Pall.
桦木科 I Betulaceae

代码 (SpCode) = BETDAH

个体数 (Individual number/20 hm^2) = 2207

最大胸径 (Max DBH) = 55.2 cm

重要值排序 (Importance value rank) = 4

落叶乔木, 5~15 m。树冠宽卵形。树皮黑褐色, 鳞块状深沟裂。小枝红褐色, 无毛。单叶互生, 卵状椭圆形, 边缘有不规则重锯齿, 近无毛, 下面密生腺点。花单性, 雌雄同株, 雄柔荑花序下垂, 雌柔荑花序直立。果序短椭圆状, 单生于短枝顶端, 直立。翅果卵形。花期5~6月, 果期8~9月。

Deciduous trees, 5-15 m tall. Crown broadly ovate. Bark black-brown, fissured. Branchlets reddish brown, glabrous. Leaves simple, alternate, petiolate, broadly ovate or elliptic, margin irregularly and doubly serrate, subglabrous, densely glandular spots below. Flowers unisexual, monoecious, male catkins pendulous and female catkins erect. Fruit inflorescences erect, oblong-cylindric, solitary at apex of short branches. Samara ovate. Fl. May-Jun., fr. Aug.-Sep..

树干　　　Trunk
摄影：林秦文　　Photo by: Lin Qinwen

叶　　　Leaves
摄影：张志翔　　Photo by: Zhang Zhixiang

花序　　　Inflorescences
摄影：沐先运　　Photo by: Mu Xianyun

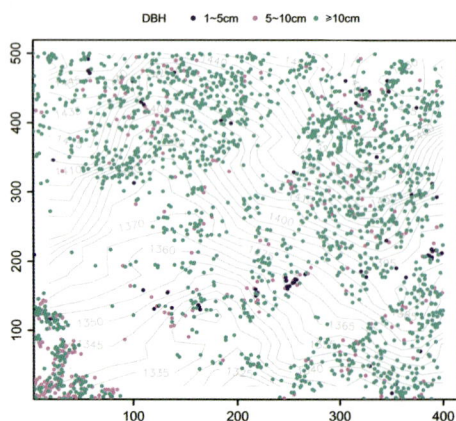

个体分布图 Distribution of individuals

径级分布表 DBH class

胸径区间 (Diameter class) (cm)	个体数 (No. of individuals in the plot)	比例 (Proportion) (%)
1~2	3	0.14
2~5	54	2.45
5~10	373	16.90
10~20	953	43.18
20~30	588	26.64
30~60	236	10.69
≥60	0	0.00

22 坚桦

Betula chinensis Maxim.
桦木科 | Betulaceae

代码 (SpCode) = BETCHI

个体数 (Individual number/20 hm²) = 12

最大胸径 (Max DBH) = 48.7 cm

重要值排序 (Importance value rank) = 40

落叶灌木或小乔木，1~5 m。树冠宽卵形。树皮暗灰色，粗糙不剥裂。叶纸质互生，卵形或广卵形，具短柄，先端钝，边缘具重锯齿。花单性，雌雄同株，雄花序生于2年生枝上；雌花序生于短枝顶端。果序近无梗，直立，近球形或椭圆形。翅果卵形，翅极窄，近于无翅。花期4~5月，果期8~9月。

Deciduous small trees, 1-5 m tall. Crown broadly ovate. Bark dark gray, coarse, not exfoliating. Leaves paper alternates, hort-petiolate, ovate or broadly ovate, apex obtuse, margin dentate-serrate. Flowers unisexual, monoecious, male flowers on biennial branches; female inflorescences on apex of short branches. Infructescence subsessile, erect, subglobose or elliptic. Samara ovate, with very narrow wings, nearly wingless. Fl. Apr.-May, fr. Aug.-Sep..

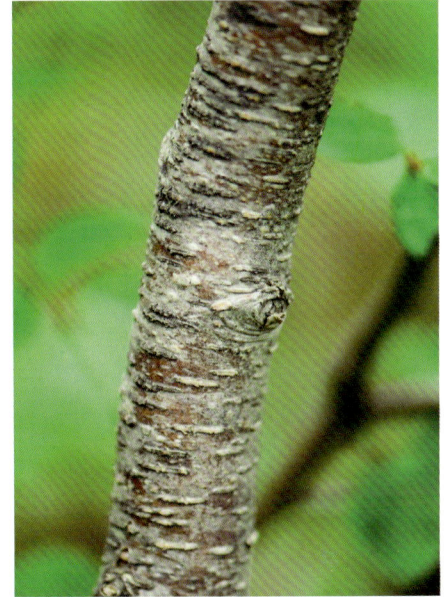

树干　　　　Trunk
摄影：汪远　　Photo by: Wang Yuan

叶与翅果　　　Leaves and samaras
摄影：林秦文　Photo by: Lin Qinwen

花序　　　Inflorescences
摄影：林秦文　Photo by: Lin Qinwen

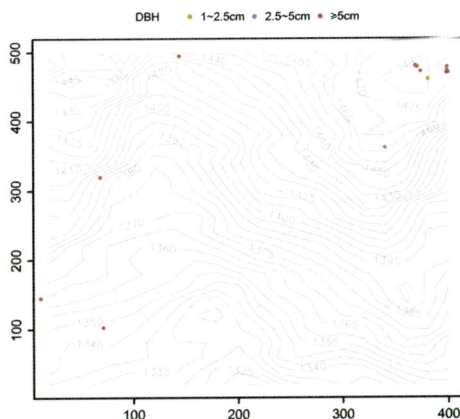

个体分布图　Distribution of individuals

径级分布表　DBH class

胸径区间 (Diameter class) (cm)	个体数 (No. of individuals in the plot)	比例 (Proportion) (%)
1~2	0	0.00
2~5	2	16.67
5~10	2	16.67
10~20	4	33.33
20~30	3	25.00
30~60	1	8.33
≥60	0	0.00

23 毛榛

Corylus sieboldiana var. *mandshurica* (Maxim.) C.K.Schneid.
桦木科 | Betulaceae

代码 (SpCode) = CORMAN

个体数 (Individual number/20 hm^2) = 4749

最大胸径 (Max DBH) = 32.1 cm

重要值排序 (Importance value rank) = 6

落叶灌木，3～4 m。树冠卵形。树皮暗灰色或灰褐色，平滑。小枝黄褐色，被长柔毛。叶互生，椭圆形，质薄，先端急尖，边缘具不规则重锯齿，中部以上具浅裂或缺刻，下面疏被短柔毛。花单性同株，雄花序2～4枚排成总状，苞鳞密被白色短柔毛；雌花序单生或簇生，腋生于雄花序上方，花柱红色。果单生或2～6枚簇生。果苞管状，较果长2～3倍，密被毛，上部浅裂。坚果卵球形，密被白绒毛。花期5～6月，果期8～9月。

Deciduous shrubs, 3-4 m tall. Crown ovate. Bark dark gray or grayish brown, smooth. Branchlets yellowish-brown, pilose. Leaves alternate, oblong, thin, base round, apex acute, margin doubly serrate, lobed above the middle, abaxially and adaxially pilose. Flowers unisexual, monoecious, male inflorescences 2-4 arranged racemose, bract scales densely white pubescent; female inflorescences solitary or fascicled, axillary above male inflorescences, style red. Fruit solitary or 2-6 clustered. Bracts forming a tubular sheath, 2-3 times longer than fruit, densely hairy, apex lobbed. Nuts ovoid-globose, white-pubescent. Fl. May-Jun., fr. Aug.-Sep..

叶　Leaves
摄影：刘冰　Photo by: Liu Bing

果　Fruits
摄影：林秦文　Photo by: Lin Qinwen

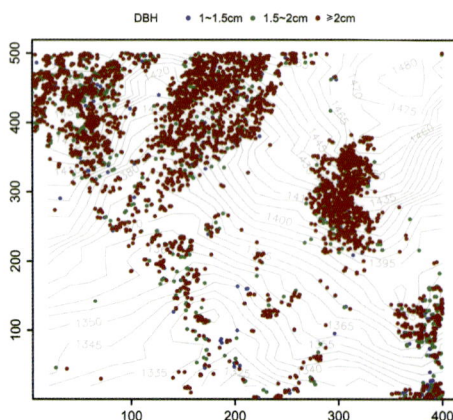

个体分布图 Distribution of individuals

径级分布表　DBH class

胸径区间 (Diameter class) (cm)	个体数 (No. of individuals in the plot)	比例 (Proportion) (%)
1～2	1640	34.53
2～5	3061	64.45
5～10	37	0.78
10～20	6	0.13
20～30	4	0.09
30～60	1	0.02
≥60	0	0.00

24 卫矛

Euonymus alatus (Thunb.) Siebold
卫矛科 | Celastraceae

代码 (SpCode) = EUOALA

个体数 (Individual number/20 hm²) = 13

最大胸径 (Max DBH) = 5.2 cm

重要值排序 (Importance value rank) = 43

落叶灌木，达3 m。树冠宽卵形。树皮灰色，木栓层发达。小枝四棱形，常具木栓翅。叶对生，纸质，窄倒卵形或椭圆形，无毛，具细锯齿。聚伞花序腋生，3～9花。花淡绿色，4数。蒴果4深裂，裂瓣长卵形，棕色带紫，假种皮橙红色。花期5～6月，果期9～10月。

Deciduous shrubs, 3 m tall. Crown broadly ovate. Bark is gray and the cork layer is developed. Twigs are four-sided and often have wooden bolt wings. Leaves are paraphyletic, papery, narrowly inverted ovate or oval, glabrous, with fine serrations. Polyumbelis inflorescence axillary, 3-9 flowers. Flowers pale green, number 4. Capsule 4 deeply lobed, lobes long ovate, brown with purple, the false seed coat is orange-red. Fl. May-Jun., fr. Sep.-Oct..

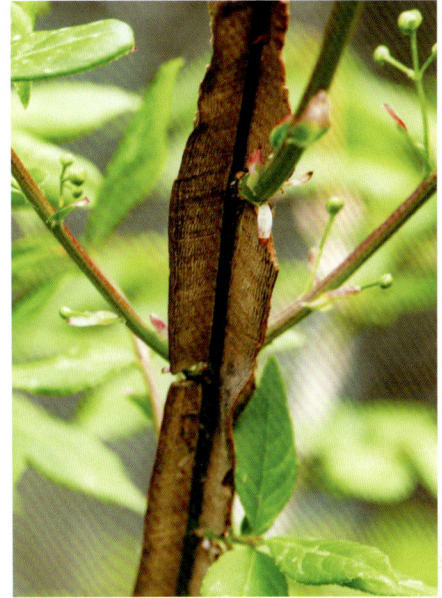

木栓质翅　　　Corky wing
摄影：张金政　　Photo by: Zhang Jinzheng

叶与花　　　Leaves and flowers
摄影：刘冰　　Photo by: Liu Bing

果　　　Fruits
摄影：林秦文　　Photo by: Lin Qinwen

个体分布图 Distribution of individuals

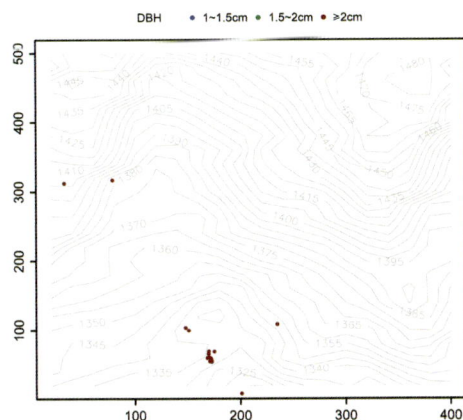

径级分布表 DBH class

胸径区间 (Diameter class) (cm)	个体数 (No. of individuals in the plot)	比例 (Proportion) (%)
1～2	1	7.69
2～5	11	84.62
5～10	1	7.69
10～20	0	0.00
20～30	0	0.00
30～60	0	0.00
≥60	0	0.00

25 山杨

Populus tremula L.
杨柳科 | Salicaceae

代码 (SpCode) = POPDAV

个体数 (Individual number/20 hm^2) = 1259

最大胸径 (Max DBH) = 46 cm

重要值排序 (Importance value rank) = 9

落叶乔木，达25 m。树冠尖塔形。树皮灰绿色或灰白色，光滑，皮孔显著。小枝圆柱形，无毛。单叶互生。叶三角状卵圆形或近圆形，无毛，边缘有波状钝齿，刚放叶时呈红色。单性异株，柔荑花序，雄花序下垂；雌花序柱头2深裂。蒴果椭圆状纺锤形，2瓣裂。花期3~4月，果期4~5月。

Deciduous trees, up to 25 m tall. Crown globose. Bark smooth, gray, lenticels conspicuous. Branchlets cylindrical, glabrous. Petiole laterally flattened. Leaves simple, alternate, leaf blade deltoid-ovate-orbicular or suborbicular, glabrous, reddish when very young, margin sinuate. Dioecious, catkin, drooping male inflorescence; female inflorescences stigma 2-lobed. Capsules ovoid-conical, 2-valved. Fl. Mar.-Apr., fr. Apr.-May.

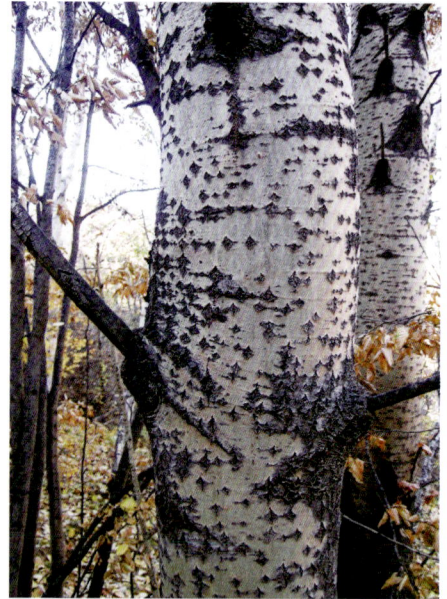

树干　　Trunk
摄影：林秦文　　Photo by: Lin Qinwen

叶背　　Leaves abaxially
摄影：张金政　　Photo by: Zhang Jinzheng

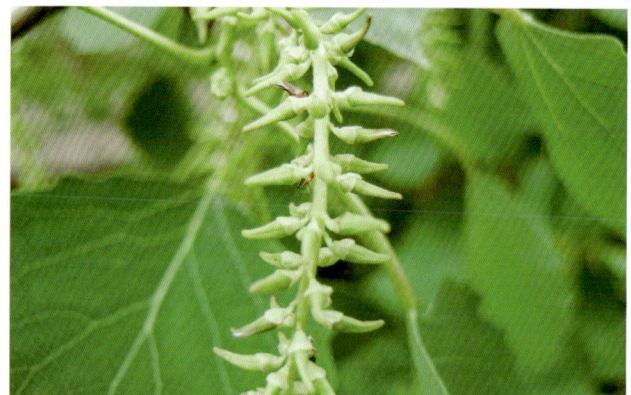

果　　Fruits
摄影：林秦文　　Photo by: Lin Qinwen

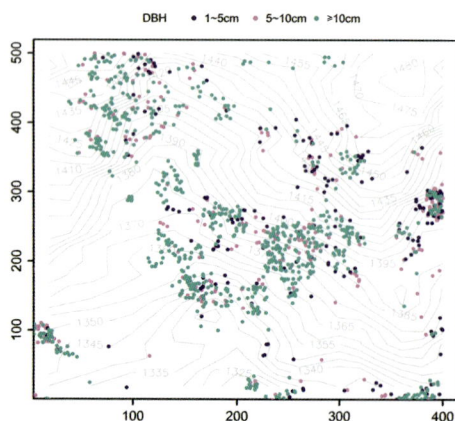

个体分布图 Distribution of individuals

径级分布表 DBH class

胸径区间 (Diameter class) (cm)	个体数 (No. of individuals in the plot)	比例 (Proportion) (%)
1~2	34	2.70
2~5	152	12.07
5~10	291	23.12
10~20	540	42.89
20~30	208	16.52
30~60	34	2.70
≥60	0	0.00

26 青杨

qīng yáng | Cathayan Poplar

Populus cathayana Rehder
杨柳科 | Salicaceae

代码 (SpCode) = POPCAT

个体数 (Individual number/20 hm^2) = 140

最大胸径 (Max DBH) = 39.5 cm

重要值排序 (Importance value rank) = 25

落叶乔木，达30 m。树冠狭卵形。树皮初光滑，灰绿色，老时暗灰色，沟裂。小枝圆柱形，无毛。叶互生，卵形至卵状长圆形，先端渐尖，基部圆形，边缘具腺圆锯齿，无毛。雄花序下垂。苞片条裂。蒴果卵圆形，3~4瓣裂。花期3~5月，果期5~7月。

Deciduous trees, up to 30 m tall. Crown narrowly ovate. Bark is initially smooth, gray-green, dark gray in old age, and cracked. Twigs cylindrical, glabrous. Leaves are alternate, ovate to ovate oblong, tapering at the apex, rounded at the base, and glandular serrated at the edges, glabrous. The male inflorescence droops. Bract clefts. The capsules are ovoid, with 3-4 lobes. Fl. Mar.-May, fr. May-Jul..

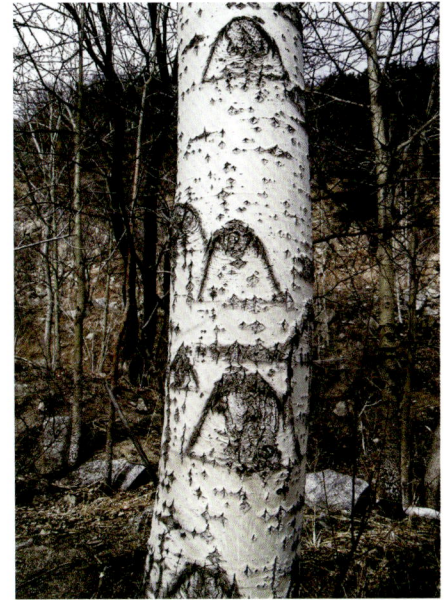

树干　　　　　　Trunk
摄影：林秦文　　Photo by: Lin Qinwen

叶　　　　　　Leaves
摄影：林秦文　　Photo by: Lin Qinwen

果　　　　　　Fruits
摄影：林秦文　　Photo by: Lin Qinwen

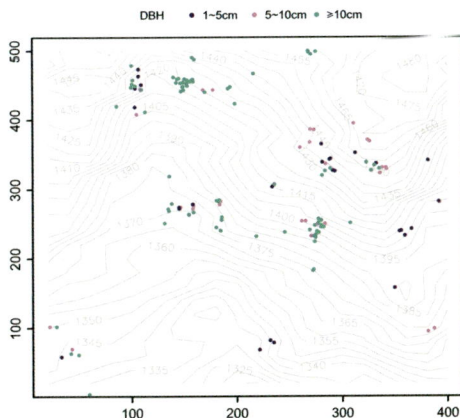

个体分布图 Distribution of individuals

径级分布表 DBH class

胸径区间 (Diameter class) (cm)	个体数 (No. of individuals in the plot)	比例 (Proportion) (%)
1~2	9	6.43
2~5	15	10.71
5~10	37	26.43
10~20	50	35.71
20~30	22	15.72
30~60	7	5.00
≥60	0	0.00

27 蒿柳

Salix schwerinii E.L.Wolf
杨柳科 | Salicaceae

代码 (SpCode) = SALVIM
个体数 (Individual number/20 hm^2) = 57
最大胸径 (Max DBH) = 55.6 cm
重要值排序 (Importance value rank) = 28

落叶灌木或乔木，达10 m。树冠宽卵形。树皮暗褐色。小枝灰色。冬芽有短柔毛。叶互生，条状披针形，近全缘，下面灰白色，密生丝状茸毛。雄花序长2～4 cm，雄蕊2，花丝离生，无毛；雌花序长3～4 cm；苞片卵形，两侧有疏长毛或短柔毛。蒴果圆形，长4～5 mm，有绢状毛。花期4～5月，果期5～6月。

Deciduous shrubs or trees, up to 10 m tall. Crown broadly ovate. Dark brown bark. Twigs gray. Winter buds have short soft hairs. Leaves are alternate, lanceolate, nearly full margin, off-white underneath and densely hairy. The male inflorescence is 2-4 cm long, stamens 2, filaments detached, glabrous; female inflorescence 3-4 cm long. Bracts are ovate with sparse or short soft hairs on both sides. Capsules are round, 4-5 mm long, with silky hairs. Fl. Apr.-May, fr. May-Jun..

叶　　　　Leaves
摄影：林秦文　　　　Photo by: Lin Qinwen

花序　　　　Inflorescences
摄影：林秦文　　　　Photo by: Lin Qinwen

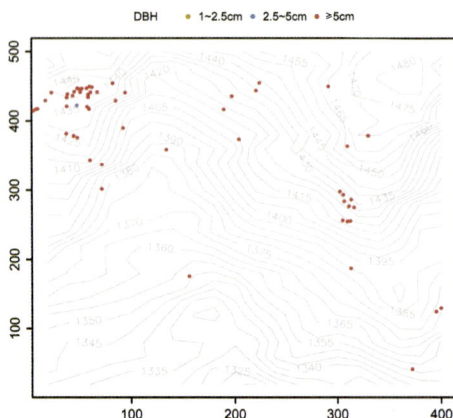

个体分布图 Distribution of individuals

径级分布表　DBH class

胸径区间 (Diameter class) (cm)	个体数 (No. of individuals in the plot)	比例 (Proportion) (%)
1～2	0	0.00
2～5	1	1.76
5～10	3	5.26
10～20	27	47.37
20～30	19	33.33
30～60	7	12.28
≥60	0	0.00

28 中国黄花柳

Salix sinica (K.S.Hao ex C.F.Fang & A.K.Skvortsov) G.H.Zhu

杨柳科 | Salicaceae

代码 (SpCode) = SALCAP

个体数 (Individual number/20 hm^2) = 3

最大胸径 (Max DBH) = 15 cm

重要值排序 (Importance value rank) = 50

落叶灌木或小乔木。叶卵状长圆形，长5～7 cm，顶部有皱褶的叶子，无毛（幼叶有柔毛），下侧被白茸毛或柔毛，有不规则缺刻或牙齿或近全缘，常轻微反卷。花先叶开放，无花序梗；雄蕊2，苞片披针形，两面密被白长毛；雌花序短圆柱形，苞片和腺体同雄花。蒴果长达9 mm，开裂后果瓣向外卷。花期4～5月，果期5～6月。

Deciduous shrubs or small trees. Leaves are ovate and oblong, 5-7 cm long, with wrinkled leaves on the top, hairless (the young leaves have soft hairs), and the underside is covered with white villi or soft hairs, with irregular missing or teeth or near the full edge, often slightly rewinded. Flowers are open with no inflorescence peduncle; stamens 2, lanceolate bracts, densely covered with long white hairs on both sides, short cylindrical female inflorescences, bracts and glands with male flowers. Capsules are up to 9 mm long and the cracking consequences are rolled outward. Fl. Apr.-May, fr. May-Jun..

叶与花序　　Leaves and inflorescences
摄影：林秦文　　Photo by: Lin Qinwen

果　　Fruits
摄影：林秦文　　Photo by: Lin Qinwen

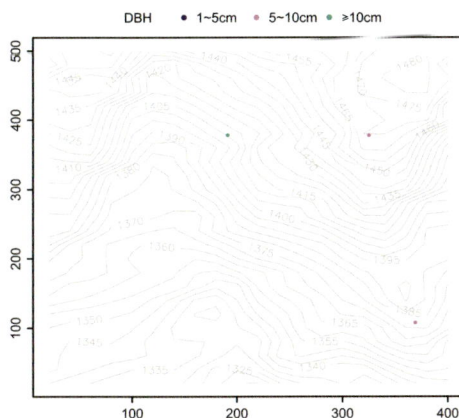

个体分布图 Distribution of individuals

径级分布表　DBH class

胸径区间 (Diameter class) (cm)	个体数 (No. of individuals in the plot)	比例 (Proportion) (%)
1～2	0	0.00
2～5	0	0.00
5～10	2	66.67
10～20	1	33.33
20～30	0	0.00
30～60	0	0.00
≥60	0	0.00

29 五角枫（色木槭） wǔ jiǎo fēng | Pentaegal Maple

Acer pictum Thunb.
槭树科 | Aceraceae

代码 (SpCode) = ACEPIC
个体数 (Individual number/20 hm^2) = 11258
最大胸径 (Max DBH) = 96 cm
重要值排序 (Importance value rank) = 2

落叶乔木，高达15～20 m。树皮粗糙，常纵裂，灰色，稀深灰色或灰褐色。小枝细瘦，无毛，具圆形皮孔。冬芽近于球形，鳞片卵形，外侧无毛，边缘具纤毛。叶纸质，基部截形或近于心脏形，叶片的外貌近于椭圆形，常5裂，裂片卵形，先端锐尖或尾状锐尖，全缘，无毛。主脉5条。圆锥状伞房花序顶生。雄花与两性花同株；萼片5，黄绿色；花瓣5，淡黄色。翅果长2～2.5 cm，小坚果压扁状，翅张开成锐角或近于钝角。花期4～5月，果期8～9月。

Deciduous trees, up to 15-20 m tall. Rough bark, often longitudinally lobed, grey, thin dark grey or grey-brown. Twigs are thin, hairless, with rounded skin holes. Winter buds are nearly spherical, ovate in scales, glabrous on the outside, and have cilia at the edges. Leaves papery, truncated at the base or nearly heart-shaped, leaf morphology nearly oval, often 5 lobes, lobes ovate, sharp apex or caudal sharp tips, full margin, glabrous. There are 5 main veins. Conical umbel inflorescence. Male flowers are the same as hermaphroditic flowers; sepals 5, yellow-green; petal 5, pale yellow. The wing fruit is 2-2.5 cm long, the small nuts are flattened, and the wings open into sharp angles or nearly obtuse angles. Fl. Apr.-May, fr. Aug.-Sep..

叶与翅果　　　　　Leaves and samaras
摄影：张志翔　　　　Photo by: Zhang Zhixiang

翅果　　　　　Samaras
摄影：叶建飞　　　　Photo by: Ye Jianfei

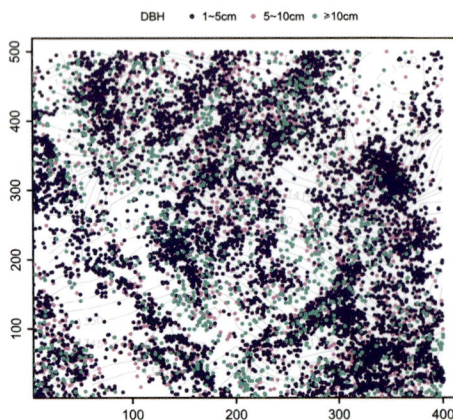

个体分布图 Distribution of individuals

径级分布表 DBH class

胸径区间 (Diameter class) (cm)	个体数 (No. of individuals in the plot)	比例 (Proportion) (%)
1～2	1546	13.73
2～5	6129	54.44
5～10	2554	22.69
10～20	859	7.63
20～30	147	1.30
30～60	22	0.20
≥60	1	0.01

30 臭椿

chòu chūn | Tree of Heaven

Ailanthus altissima (Mill.) Swingle
苦木科 | Simaroubaceae

代码 (SpCode) = AILALT

个体数 (Individual number/20 hm^2) = 3

最大胸径 (Max DBH) = 4 cm

重要值排序 (Importance value rank) = 51

落叶乔木，达20 m。树冠卵形。树皮平滑，具纵浅裂纹，暗褐色。小枝赤褐色，被疏柔毛。单数羽状复叶互生，大型，小叶13～25，卵状披针形，近基部具1臭腺体。叶柄粗壮。圆锥花序顶生；花杂性，白色带绿。翅果矩圆状椭圆形，黄绿色或紫红色。花期10～11月，果期翌年1～3月。

Deciduous trees, up to 20 m tall. Crown ovate. Bark is smooth, with longitudinal and shallow cracks, dark brown. Twigs are russet and covered with soft hairs. Singular pinnate compound leaves are alternate, large, leaflets 13-25, ovate lanceolate, with 1 odor gland near the base. The petiole is stout. Conical inflorescence apical; floral heterogeneity, white with green. The winged fruit is rectangular and oval, yellow-green or purple-red. Fl. Oct.-Nov., fr. Jan.-Mar. of next year.

植株　　Whole plant
摄影：林秦文　Photo by: Lin Qinwen

叶　　Leaves
摄影：林秦文　Photo by: Lin Qinwen

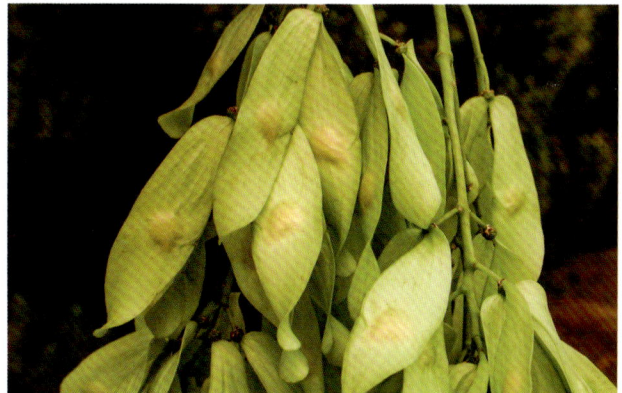

果序　　Infructescences
摄影：林秦文　Photo by: Lin Qinwen

径级分布表 DBH class

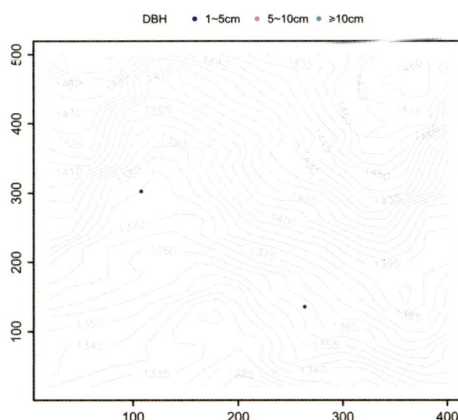

个体分布图 Distribution of individuals

胸径区间 (Diameter class) (cm)	个体数 (No. of individuals in the plot)	比例 (Proportion) (%)
1～2	1	33.30
2～5	2	66.70
5～10	0	0.00
10～20	0	0.00
20～30	0	0.00
30～60	0	0.00
≥60	0	0.00

31 辽椴（糠椴、大叶椴） liáo duàn | Manchurian Liden

Tilia mandshurica Rupr. et Maxim.
椴树科 | Tiliaceae

代码 (SpCode) = TILMAN

个体数 (Individual number/20 hm^2) = 418

最大胸径 (Max DBH) = 36.8 cm

重要值排序 (Importance value rank) = 19

落叶乔木，达20 m。树冠宽卵形。树皮暗灰色，纵裂。小枝被白色星状毛。单叶互生，纸质，卵圆形，下面密被灰色星状茸毛，侧脉5～7对，边缘具尖锯齿。聚伞花序，花6～12朵，有毛。苞片窄长圆形，苞片下部1/3与花序梗合生，萼片5，花瓣5，退化雄蕊5，花瓣状。核果球形，有5条不明显的棱，被毛。花期6～7月，果期8～9月。

Deciduous trees, up to 20 m tall. Crown broadly ovate. Bark dark gray, longitudinally fissured. Branchlets white-stellatetomentose. Leaves simple, alternate, ovate orbicular, densely gray stellatetomentose abaxially, lateral veins 5-7 pairs, margin dentate. Inflorescences cyme, 6-12-flowered, hairy. Bract narrow oblong, lower 1/3 to 1/2 adnate to peduncle. Sepals 5, petals 5, staminodes 5, petals shape. Fruit globular, with 5 inconspicuous edges, hairy. Fl. Jun.-Jul., fr. Aug.-Sep..

叶与花序　Leaves and inflorescences
摄影：林秦文　Photo by: Lin Qinwen

花序　Inflorescences
摄影：林秦文　Photo by: Lin Qinwen

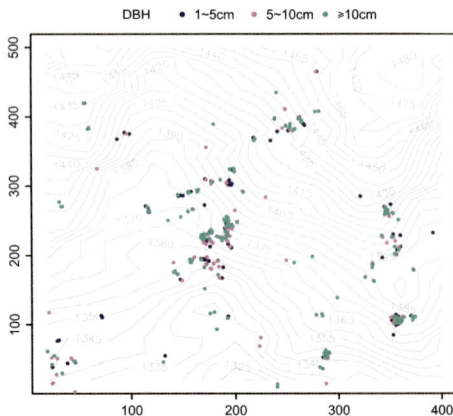

个体分布图 Distribution of individuals

径级分布表　DBH class

胸径区间 (Diameter class) (cm)	个体数 (No. of individuals in the plot)	比例 (Proportion) (%)
1～2	9	2.15
2～5	107	25.60
5～10	126	30.14
10～20	141	33.73
20～30	34	8.14
30～60	1	0.24
≥60	0	0.00

32 蒙椴（小叶椴）

méng duàn | Mongol Linden

Tilia mongolica Maxim.
椴树科 | Tiliaceae

代码 (SpCode) = TILMON

个体数 (Individual number/20 hm^2) = 415

最大胸径 (Max DBH) = 54.9 cm

重要值排序 (Importance value rank) = 15

落叶乔木，达10 m。树冠宽卵形。树皮灰褐色，不规则薄片状脱落。单叶互生，纸质，阔卵形或圆形，常3裂，下面仅脉腋内有毛丛，侧脉4～5对，边缘有粗锯齿，顶端锐尖。花6～12朵组成聚伞花序，苞片下半部与花序梗合生。萼片5，披针形；花瓣5，花瓣状。核果倒卵形，外被绒毛。花期6～7月，果期9～11月。

Deciduous trees, up to 10 m tall. Crown broadly ovate. Bark grayish, irregularly exfoliate. Leaves simple, alternate, papery, broad ovate or orbicular, usually 3-lobed, abaxially hairy only in axils of veins, lateral veins 4-5 pairs, margin roughly serrate, apex acuminate. Inflorescences cymose, 6-12-flowered, lower half adnate with peduncle. Sepals 5, lanceolate; petals 5, petals shape. Fruits obovate, hairy on the surface. Fl. Jun.-Jul., fr. Sep.-Nov..

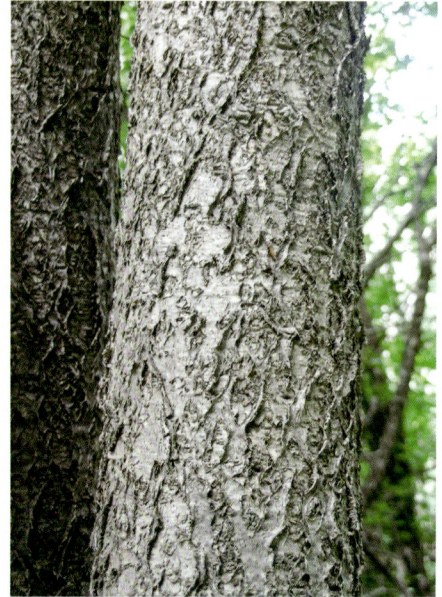

树干	Trunk

摄影：林秦文　　Photo by: Lin Qinwen

叶与花序	Leaves and inflorescences

摄影：林秦文　　Photo by: Lin Qinwen

花	Flowers

摄影：林秦文　　Photo by: Lin Qinwen

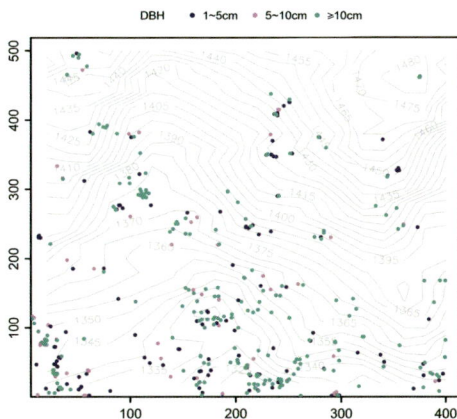

个体分布图 Distribution of individuals

径级分布表　DBH class

胸径区间 (Diameter class) (cm)	个体数 (No. of individuals in the plot)	比例 (Proportion) (%)
1～2	9	2.17
2～5	96	23.13
5～10	94	22.65
10～20	148	35.66
20～30	53	12.77
30～60	15	3.62
≥60	0	0.00

33 沙梾

Cornus bretschneideri L. Henry
山茱萸科 | Cornaceae

代码 (SpCode) = CORBRE

个体数 (Individual number/20 hm^2) = 101

最大胸径 (Max DBH) = 5.9 cm

重要值排序 (Importance value rank) = 26

灌木或小乔木，高1～6 m。树皮紫红色。幼枝圆柱形，带红色，有稀疏的贴生灰白色短柔毛。叶对生，纸质，卵形，先端突尖或短渐尖，下面灰白色，密被不明显的乳头状突起及白色贴生的短柔毛。伞房状聚伞花序顶生。花白。花瓣舌状长卵形。核果蓝黑色至黑色，近于球形，密被贴生短柔毛。花期6～7月，果期8～9月。

Shrubs or small trees, 1-6 m tall. Bark purplish red. Young branches reddish, terete, sparsely pubescent with grayish white trichomes. Leaves opposite, leaf blade ovate, papery, abaxially grayish white or glaucous, densely papillose and pubescent with appressed white trichomes or occasionally with dense yellowish brown curly trichomes, apex cuspidate to acuminate. Corymbose cymes. Flowers white. Petals ligulate to ovate. Fruit bluish black or black, subglobose. Fl. Jun.-Jul., fr. Aug.-Sep..

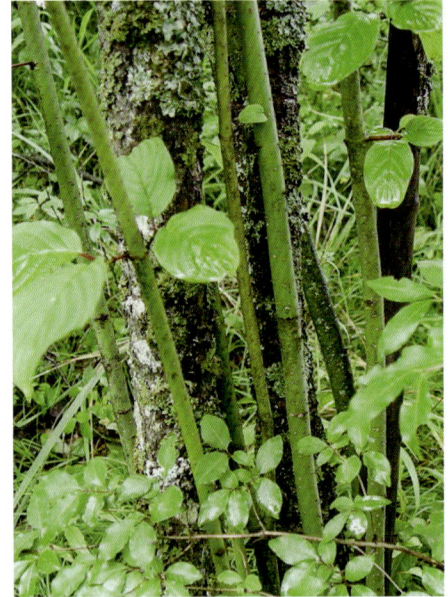

树干　　　　　　Trunk
摄影：肖翠　　　Photo by: Xiao Cui

叶背　　　　　　Leaves abaxially
摄影：林秦文　　Photo by: Lin Qinwen

花序　　　　　　Inflorescences
摄影：林秦文　　Photo by: Lin Qinwen

径级分布表　DBH class

胸径区间 (Diameter class) (cm)	个体数 (No. of individuals in the plot)	比例 (Proportion) (%)
1～2	33	32.67
2～5	66	65.35
5～10	2	1.98
10～20	0	0.00
20～30	0	0.00
30～60	0	0.00
≥60	0	0.00

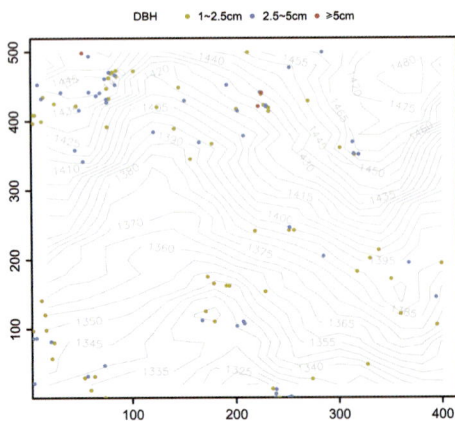

个体分布图　Distribution of individuals

34 钩齿溲疏

Deutzia baroniana Diels
(*Deutzia hamata* Koehnc var. *baroniana* (Diels) Zaikonn)
绣球花科 | Hydrangeaceae

代码 (SpCode) = DEUHAM

个体数 (Individual number/20 hm^2) = 31

最大胸径 (Max DBH) = 4.7 cm

重要值排序 (Importance value rank) = 35

灌木，高0.3～1 m。花枝具2～4叶，浅褐色，被星状毛。叶纸质，卵状菱形或卵状椭圆形，先端急尖，边缘具不整齐或大小相间锯齿，上面疏被4～5辐线星状毛。叶柄疏被星状毛。聚伞花序具2～3花或花单生。花瓣白色，倒卵状长圆形，先端圆形，下部收狭，外面被星状毛，花蕾时内向镊合状排列。花丝先端2齿，齿平展或下弯成钩状。蒴果半球形，密被星状毛。花期4～5月，果期9～10月。

Shrubs, 0.3-1 m tall. Flowering branchlets brownish, 2-4 leaved, stellate hairy. Leaves are papery, ovate diamond-shaped or ovate-oval, sharply pointed at the apex, irregular or serrated at the edges, sparsely covered with 4-5 spoke stellate hairs. Petioles are sparsely covered with stellate hairs. Polyumbelis inflorescences have 2-3 flowers or solitary flowers. Petals are white, inverted ovate oblong, the apex is rounded, the lower part is narrowed, the outer part is covered with stellate hairs, and the buds are arranged inwardly inward- and tweezered. The filament has 2 teeth at the apex, the teeth are flattened or bent down into a hook. Capsules are hemispherical, densely covered with stellate hairs. Fl. Apr.-May, fr. Sep.-Oct..

叶背　　　　　　Leaves abaxially
摄影：林秦文　　Photo by: Lin Qinwen

花　　　　　　Flower
摄影：林秦文　　Photo by: Lin Qinwen

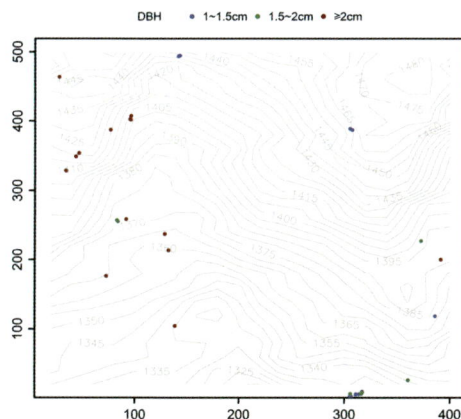

个体分布图 Distribution of individuals

径级分布表 DBH class

胸径区间 (Diameter class) (cm)	个体数 (No. of individuals in the plot)	比例 (Proportion) (%)
1～2	17	54.84
2～5	14	45.16
5～10	0	0.00
10～20	0	0.00
20～30	0	0.00
30～60	0	0.00
≥60	0	0.00

35 小花溲疏

Deutzia parviflora Bunge
绣球花科 | Hydrangeaceae

代码 (SpCode) = DEUPAR
个体数 (Individual number/20 hm^2) = 1106
最大胸径 (Max DBH) = 26 cm
重要值排序 (Importance value rank) = 14

落叶灌木，1~2 m。树冠卵形。树皮灰白色，块状剥落。小枝疏被星状毛。叶对生，卵形或狭卵形，边缘具小锯齿，两面疏被星状毛。花序伞房状，具多数花。花萼密生星状毛，裂片5，宽卵形。花瓣5，覆瓦状排列，白色，倒卵形，长约6 mm。雄蕊10，花丝无齿或上部具短钝齿。子房下位，花柱3。蒴果半球形，直径2~3 mm。花期5~6月，果期8~10月。

Deciduous shrubs, 1-2 m tall. Crown ovate. Bark is off-white and flaky. Twigs are sparsely covered with stellate hairs. Leaves are opposite, ovate or narrowly ovate, with small serrated margins and sparsely covered with stellate hairs on both sides. The inflorescence is umbel-shaped and has most flowers. Calyx densely stellate hairs, lobes 5, broadly ovate. Petals 5, tile-like arrangement, white, inverted ovate, about 6 mm long. Stamens 10, filaments toothless or with short blunt teeth on the upper part. Lower sub-chamber, pillar 3. Capsules hemispherical, 2-3 mm in diameter. Fl. May-Jun., fr. Aug.-Oct..

叶背　　Leaves abaxially
摄影：林秦文　　Photo by: Lin Qinwen

花序　　Inflorescences
摄影：林秦文　　Photo by: Lin Qinwen

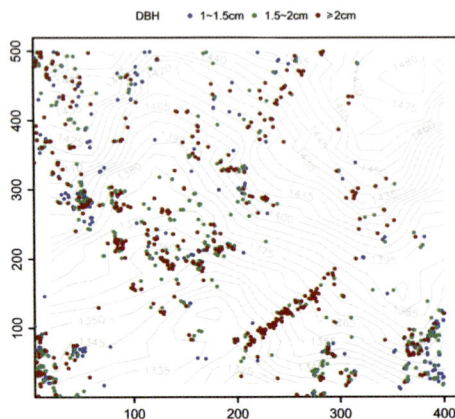

个体分布图 Distribution of individuals

径级分布表 DBH class

胸径区间 (Diameter class) (cm)	个体数 (No. of individuals in the plot)	比例 (Proportion) (%)
1~2	745	67.36
2~5	350	31.65
5~10	6	0.54
10~20	4	0.36
20~30	1	0.09
30~60	0	0.00
≥60	0	0.00

36　太平花

Philadelphus pekinensis Rupr.
绣球花科 | Hydrangeaceae

代码 (SpCode) = PHIPEK
个体数 (Individual number/20 hm^2) = 19
最大胸径 (Max DBH) = 5.5 cm
重要值排序 (Importance value rank) = 33

落叶灌木，达2 m。树冠宽卵形。树皮灰白色。小枝无毛，红褐色。叶对生，卵形，先端渐尖，边缘有小锯齿，两面无毛。聚伞状总状花序，花序具5～9花。花序轴和花梗都无毛。萼筒无毛，裂片4，三角状卵形，外面无毛；花瓣4，白色，倒卵形；雄蕊多数；子房下位，4室，花柱上部4裂，柱头近匙形。蒴果球状倒圆锥形。花期4～6月，果期8～10月。

Deciduous shrubs, up to 2 m tall. Crown broadly ovate. Bark is off-white. Twigs are glabrous, reddish-brown. Leaves are opposite, ovate, tapering at the apex, with small serrations at the edges, and hairless on both sides. Polyumbelous inflorescences with 5-9 flowers. Both the inflorescence axis and the peduncle are glabrous. Calyx glabrous, lobe 4, triangular ovate, glabrous on the outside; petals 4, white, inverted ovate; stamens majority; lower ovary, 4 loculars, 4 splits in the upper part of the peduncle, nearly spoon-shaped stigma. Capsules are spherical inverted conical. Fl. Apr.-Jun., fr. Aug.-Oct..

植株　　Whole plant
摄影：林秦文　　Photo by: Lin Qinwen

枝干　　Twigs
摄影：张金政　　Photo by: Zhang Jinzheng

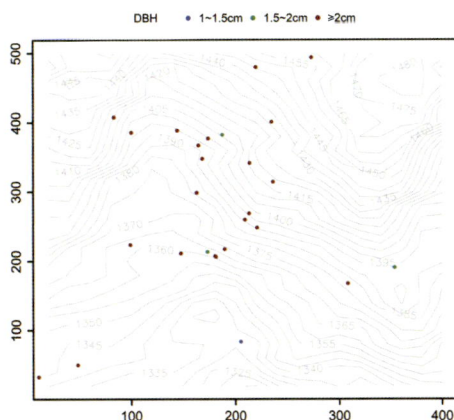

个体分布图　Distribution of individuals

径级分布表 DBH class

胸径区间 (Diameter class) (cm)	个体数 (No. of individuals in the plot)	比例 (Proportion) (%)
1～2	3	15.79
2～5	15	78.95
5～10	1	5.26
10～20	0	0.00
20～30	0	0.00
30～60	0	0.00
≥60	0	0.00

37 东陵绣球

Hydrangea bretschneideri Dippel
绣球花科 | Hydrangeaceae

代码 (SpCode) = HYDBRE
个体数 (Individual number/20 hm^2) = 135
最大胸径 (Max DBH) = 14.3 cm
重要值排序 (Importance value rank) = 24

落叶灌木，1～3 m。树冠宽卵形。树皮较薄，薄片状剥落。小枝栗褐色。叶对生，纸质，卵形至长椭圆形，先端渐尖，边缘具粗齿，下面密被柔毛。伞房状聚伞花序。不育花萼片4，大型，花瓣状，白色。孕性花萼筒杯状，萼齿三角形。花瓣白色，卵状披针形或长圆形；雄蕊10；子房半下位，花柱3，基部连合，柱头近头状。蒴果卵球形，突出部分圆锥形。花期6～7月，果期9～10月。

Deciduous shrubs, 1-3 m tall. Crown broadly ovate. Bark is thin and flaky and peeling. Twigs chestnut brown. Leaves are opposite, papery, ovate to oblong-oval, tapering at the apex, coarsely toothed at the edges, and densely hairy underneath. Umbelliferous polyumpses. Sterile calyx 4, large, petal-like, white. Pregnant calyx cup-shaped, calyx triangular. Petals are white, ovate lanceolate or oblong; stamens 10; ovary semi-inferior position, column 3, base contiguous, stigma near head. Capsules are ovoid and protruding partially conical. Fl. Jun.-Jul., fr. Sep.-Oct..

叶背　Leaves abaxially
摄影：张鑫　Photo by: Zhang Xin

花序　Inflorescences
摄影：林秦文　Photo by: Lin Qinwen

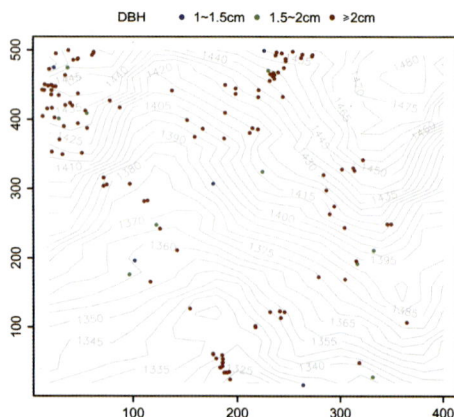

个体分布图　Distribution of individuals

径级分布表　DBH class

胸径区间 (Diameter class) (cm)	个体数 (No. of individuals in the plot)	比例 (Proportion) (%)
1～2	16	11.85
2～5	57	42.22
5～10	52	38.52
10～20	10	7.41
20～30	0	0.00
30～60	0	0.00
≥60	0	0.00

38 照山白

Rhododendron micranthum Turcz.

杜鹃花科 | Ericaceae

代码 (SpCode) = RHOMIC

个体数 (Individual number/20 hm^2) = 1256

最大胸径 (Max DBH) = 35 cm

重要值排序 (Importance value rank) = 7

常绿灌木，达2.5 m。树冠卵形。树皮灰棕褐色，平滑。小枝被鳞片及细柔毛；叶互生，近革质，倒披针形至披针形，顶端钝，基部狭楔形，下面被棕色鳞片。总状花序顶生，有花10～28朵，花密集，花小，乳白色。花冠钟状，花裂片5。蒴果长圆形，被疏鳞片。花期5～6月，果期8～11月。

Evergreen shrubs, up to 2.5 m tall. Crown ovate. Bark grey-brown, smooth. Twigs covered with scales and fine soft hairs; alternating, nearly leathery, inverted lanceolate to lanceolate, blunt at the apex, narrowly wedge-shaped at the base, and brown scales below. The inflorescence is apical, with 10-28 flowers, dense flowers, small flowers, milky white. Corolla bell-shaped, flower lobes 5. Capsules are oblong and sparsely scaled. Fl. May-Jun., fr. Aug.-Nov..

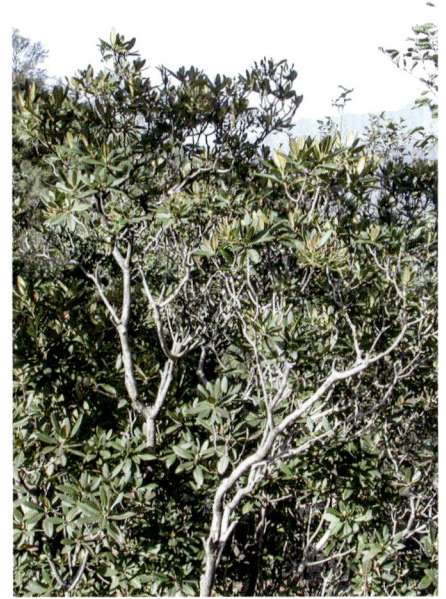

植株　Whole plant
摄影：林秦文　Photo by: Lin Qinwen

叶背　Leaves abaxially
摄影：李晓东　Photo by: Li Xiaodong

花序　Inflorescences
摄影：林秦文　Photo by: Lin Qinwen

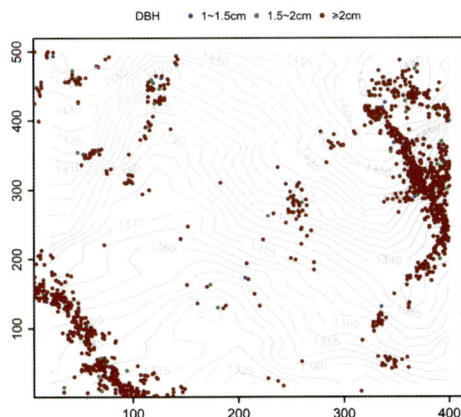

个体分布图 Distribution of individuals

径级分布表　DBH class

胸径区间 (Diameter class) (cm)	个体数 (No. of individuals in the plot)	比例 (Proportion) (%)
1～2	146	11.62
2～5	1060	84.40
5～10	46	3.66
10～20	2	0.16
20～30	1	0.08
30～60	1	0.08
≥60	0	0.00

39 迎红杜鹃

yíng hóng dù juān | Wheeldon Pink

Rhododendron mucronulatum Turcz.
杜鹃花科 | Ericaceae

代码 (SpCode) = RHOMUC

个体数 (Individual number/20 hm²) = 1000

最大胸径 (Max DBH) = 15 cm

重要值排序 (Importance value rank) = 13

落叶灌木，1~2 m。树冠宽卵形。树皮灰白色，平滑。小枝细长，疏生鳞片。叶互生，质薄，椭圆形或椭圆状披针形，全缘或有细圆齿，下面被褐色鳞片。花序腋生枝顶或假顶生，1~3花，先叶开放，伞形着生。花冠宽漏斗状，淡红紫色。蒴果长圆形。花期4~6月，果期5~7月。

Deciduous shrubs, 1-2 m tall. Crown broadly ovate. Bark is off-white and smooth. Slender twigs, sparse scales. Leaves are alternate, thin, oval or oval lanceolate, fully margined or with fine rounded teeth, and underneath them are brown scales. The inflorescence is axillary or pseudo-apical, 1-3 flowers, the first leaves are open, and the umbel is born. The corolla is wide funnel-shaped, reddish-purple. The capsule is oblong. Fl. Apr.-Jun., fr. May-Jul..

植株　　Whole plant
摄影：林秦文　Photo by: Lin Qinwen

叶背　　Leaves abaxially
摄影：张金政　Photo by: Zhang Jinzheng

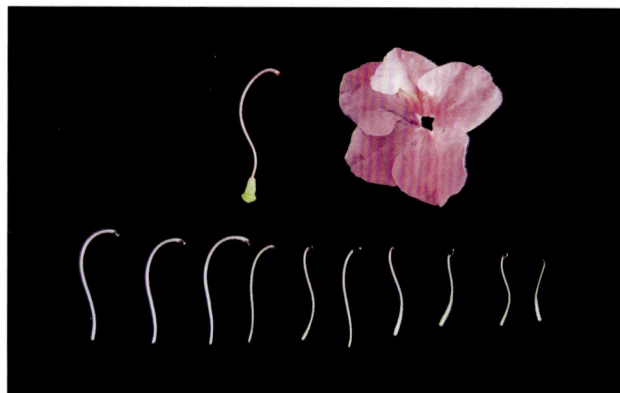

花　　Flower
摄影：汪远　Photo by: Wang Yuan

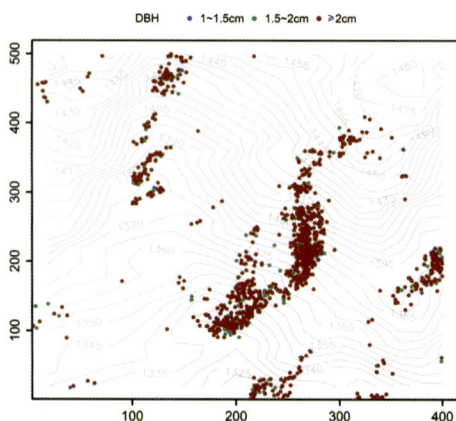

个体分布图 Distribution of individuals

径级分布表 DBH class

胸径区间 (Diameter class) (cm)	个体数 (No. of individuals in the plot)	比例 (Proportion) (%)
1~2	204	20.40
2~5	795	79.50
5~10	0	0.00
10~20	1	0.10
20~30	0	0.00
30~60	0	0.00
≥60	0	0.00

40　花曲柳（大叶白蜡）　　　　huā qū liǔ | Chinese Ash

Fraxinus chinensis subsp. *rhynchophylla* (Hance) A.E.Murray
木犀科 | Oleaceae

代码 (SpCode) = FRACHI

个体数 (Individual number/20 hm^2) = 1590

最大胸径 (Max DBH) = 46 cm

重要值排序 (Importance value rank) = 10

落叶乔木。树皮灰褐色，光滑，老时浅裂。奇数羽状复叶，对生，小叶5～7枚，阔卵形或卵状披针形，边缘有不明显的锯齿，顶生小叶明显大于侧生小叶。圆锥花序生于当年生枝顶，花密集，黄绿色，雄花与两性花异株。花萼浅杯状，无花冠；雄蕊2枚；雌蕊具短花柱，柱头2深裂。翅果狭倒披针形。花期4～5月，果期9～10月。

Deciduous trees. Bark is grayish brown, smooth and shallowly cracked in old age. Odd pinnate compound leaves, paraphytic, 5-7 leaflets, broadly ovate or ovate lanceolate, with inconspicuous serrations at the edges, apical leaflets significantly larger than lateral leaflets. Conical inflorescences are borne at the top of the branches of the year, the flowers are dense, yellow-green, and the male flowers are heterogeneous with hermaphroditic flowers. Calyx shallow cup-shaped, no corolla; 2 stamens; the pistils have short flower pillars with 2 deep lobes at the stigma. Winged fruit is narrowly inverted lanceolate. Fl. Apr.-May, fr. Sep.-Oct..

叶　　　　Leaves
摄影：林秦文　　　　Photo by: Lin Qinwen

翅果　　　　Samaras
摄影：林秦文　　　　Photo by: Lin Qinwen

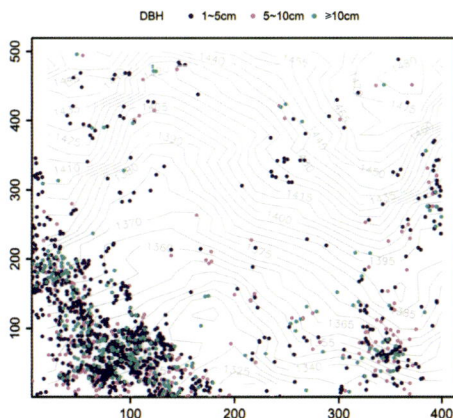

个体分布图　Distribution of individuals

径级分布表　DBH class

胸径区间 (Diameter class) (cm)	个体数 (No. of individuals in the plot)	比例 (Proportion) (%)
1～2	108	6.79
2～5	721	45.35
5～10	519	32.65
10～20	233	14.65
20～30	8	0.50
30～60	1	0.06
≥60	0	0.00

41 北京丁香

Syringa reticulata subsp. *pekinensis* (Rupr.) P. S. Green & M. C. Chang
木犀科 I Oleaceae

代码 (SpCode) = SYRRET

个体数 (Individual number/20 hm^2) = 282

最大胸径 (Max DBH) = 26.8 cm

重要值排序 (Importance value rank) = 22

大灌木或落叶乔木，4～10 m。树冠宽卵形。树皮紫灰褐色，具细裂纹。小枝灰褐色，无毛，疏生皮孔。叶对生，厚纸质，宽卵形至长圆状披针形，全缘，无毛，先端常尾尖。圆锥花序发自侧芽；花萼小，萼齿钝至截平；花冠辐状，白色，长4～5 mm，裂片4，卵形；雄蕊2枚，花药黄色；子房2室。蒴果长椭圆形，长1.5～2 cm，熟后开裂。花期6～7月，果期8～10月。

Large shrubs or deciduous trees, 4-10 m tall. Crown broadly ovate. Bark is purple and grayish brown with fine cracks. Twigs gray-brown, glabrous, sparse skin holes. Leaves are opposite, thick papery, broadly ovate to oblong lanceolate, fully margined, glabrous, often pointed at the apex. The conical inflorescence originates from lateral buds; calyxes are small, and the calyx teeth are blunt to truncated; corolla spoke, white, 4-5 mm long, lobes 4, ovate; 2 stamens, anther yellow; 2 rooms in the ovary. Capsules are oblong, 1.5-2 cm long, cracking after ripening. Fl. Jun.-Jul., fr. Aug.-Oct..

叶与花序　Leaves and inflorescences
摄影：叶建飞　Photo by: Ye Jianfei

果序　Infructescences
摄影：林秦文　Photo by: Lin Qinwen

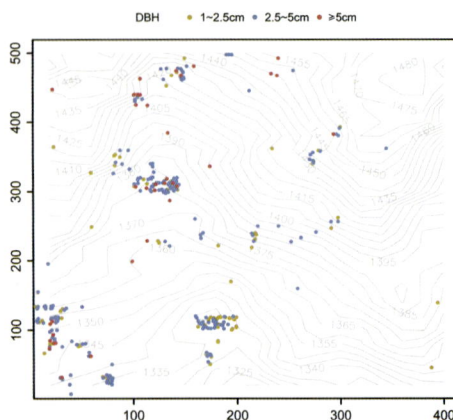

个体分布图 Distribution of individuals

径级分布表　DBH class

胸径区间 (Diameter class) (cm)	个体数 (No. of individuals in the plot)	比例 (Proportion) (%)
1～2	51	18.09
2～5	203	71.99
5～10	24	8.51
10～20	3	1.06
20～30	1	0.35
30～60	0	0.00
≥60	0	0.00

42　巧玲花

qiǎo líng huā | Pubescent Lilac

Syringa pubescens Turcz.
木犀科 I Oleaceae

代码 (SpCode) = SYRPUB

个体数 (Individual number/20 hm^2) = 4462

最大胸径 (Max DBH) = 32.3 cm

重要值排序 (Importance value rank) = 5

灌木。叶对生，卵形或菱状卵形，上面深绿色，无毛，下面淡绿色，叶脉基部密被柔毛，叶缘具睫毛。圆锥花序直立，由侧芽抽生。花冠初开时紫色，后渐进白色，花冠管细长，裂片4，开展或反折；雄蕊2，花药紫色。蒴果长椭圆形，具瘤状凸起。花期5～6月，果期6～8月。

Shrubs. Leaves are opposite, ovate or diamond-shaped, dark green on the top, glabrous, pale green below, densely hairy at the base of the leaf veins, and eyelashes on the leaf margins. Conical inflorescence is erect and drawn from lateral buds. Corollas are purple at first opening, and then gradually white, the corolla tube is slender, lobed 4, unfolded or folded; stamens 2, anthers purple. Capsules are oblong-oval with noomatous bumps. Fl. May-Jun., fr. Jun.-Aug..

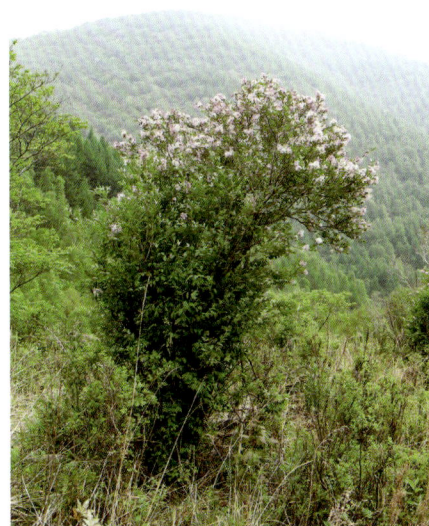

植株　　Whole plant
摄影：林秦文　　Photo by: Lin Qinwen

叶与花序　　Leaves and inflorescences
摄影：林秦文　　Photo by: Lin Qinwen

果序　　Infructescences
摄影：林秦文　　Photo by: Lin Qinwen

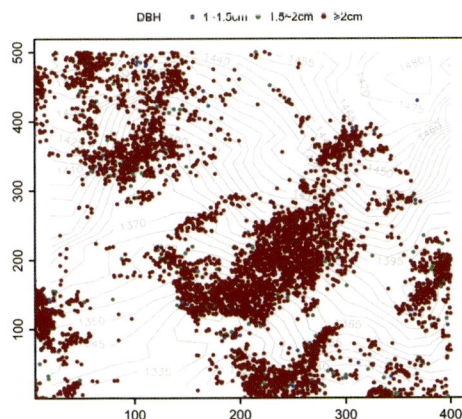

个体分布图 Distribution of individuals

径级分布表　DBH class

胸径区间 (Diameter class) (cm)	个体数 (No. of individuals in the plot)	比例 (Proportion) (%)
1～2	501	11.22
2～5	3648	81.76
5～10	302	6.77
10～20	8	0.18
20～30	0	0.00
30～60	3	0.07
≥60	0	0.00

43 六道木

Zabelia biflora (Turcz.) Makino ex Hisauti & H.Hara
忍冬科 | Caprifoliaceae

代码 (SpCode) = ABEBIF
个体数 (Individual number/20 hm^2) = 4506
最大胸径 (Max DBH) = 63 cm
重要值排序 (Importance value rank) = 3

落叶灌木，1～3 m。树冠卵形。树皮具棱，黑褐色。小枝纤细，被毛。叶对生，矩圆状披针形，顶端尖，基部钝，全缘或浅裂，两面被柔毛。花单生叶腋，无总梗，具苞片；花冠白色，漏斗形，4裂，被毛。瘦果矩圆形，具纵棱。种子近圆柱形，种皮膜质。花期5～6月，果期8～9月。

Deciduous shrubs, 1-3 m tall. Crown ovate. Bark is ribbed, black-brown. Slender twigs, covered with hairs. Leaves are paraphyletic, rectangular lanceolate, pointed at the apex, blunt at the base, fully margined or shallowly lobed, and covered with soft hairs on both sides. Flowers are solitary leaf axillary, without total stems, with bracts; corolla white, funnel-shaped, 4 lobed, industibility. Fruits are rounded, with longitudinal edges. Seeds are nearly cylindrical, membranous in the seed coat. Fl. May-Jun., fr. Aug.-Sep..

枝　　　　　Branch
摄影：林秦文　　Photo by: Lin Qinwen

叶和花　　　　　Leaves and flower
摄影：刘冰　　Photo by: Liu Bing

果　　　　　Fruits
摄影：王小然　　Photo by: Wang Xiaoran

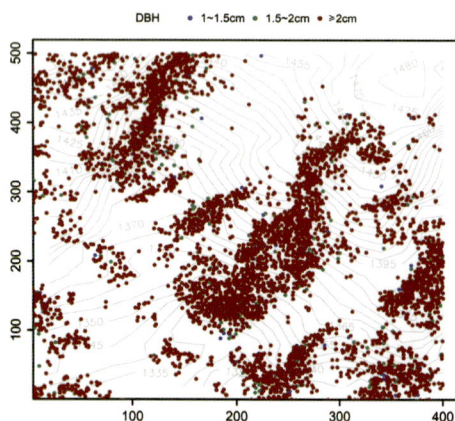

个体分布图 Distribution of individuals

径级分布表 DBH class

胸径区间 (Diameter class) (cm)	个体数 (No. of individuals in the plot)	比例 (Proportion) (%)
1～2	410	9.10
2～5	3406	75.59
5～10	682	15.14
10～20	5	0.11
20～30	1	0.02
30～60	1	0.02
≥60	1	0.02

44 金花忍冬　　jīn huā rěn dōng | Coralline Honeysuckle

Lonicera chrysantha Turcz. ex Ledeb.
忍冬科 | Caprifoliaceae

代码 (SpCode) = LONCHR

个体数 (Individual number/20 hm^2) = 58

最大胸径 (Max DBH) = 11.9 cm

重要值排序 (Importance value rank) = 29

落叶灌木，达2 m。树冠卵形。树皮暗灰色。小枝纤细，被长毛。叶对生，菱状卵形至菱形状披针形，全缘，顶端渐尖，两面被长柔毛。双花总花梗腋生，直立。花冠先白色后黄色，唇形，花冠筒3倍短于唇瓣。浆果近球形，红色。花期5～6月，果期8～9月。

Deciduous shrubs, up to 2 m tall. Crown ovate. Bark is dark gray. Twigs are slender and covered with long hairs. Leaves are opposite, rhomboid ovate to rhomboid lanceolate, full margin, tapering at the apex, and covered with long soft hairs on both sides. Double-flowered total peduncle axillary, erect. Corollas are first white and then yellow, lip-shaped, and the corolla tube is 3 times shorter than the lip flap. Berries are nearly spherical, red. Fl. May-Jun., fr. Aug.-Sep..

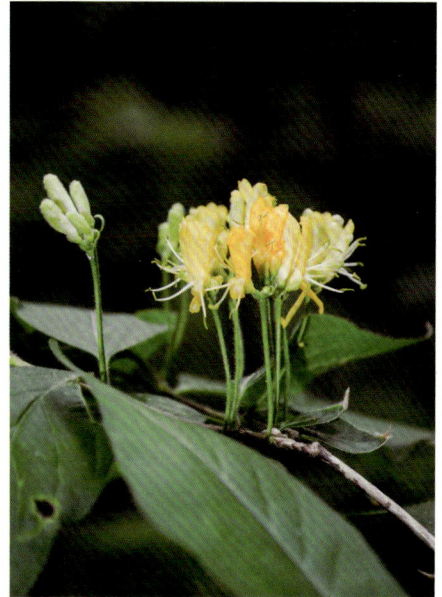

花序　Inflorescences
摄影：叶建飞　Photo by: Ye Jianfei

叶背　Leaves abaxially
摄影：刘晓　Photo by: Liu Xiao

果序　Infructescences
摄影：刘冰　Photo by: Liu Bing

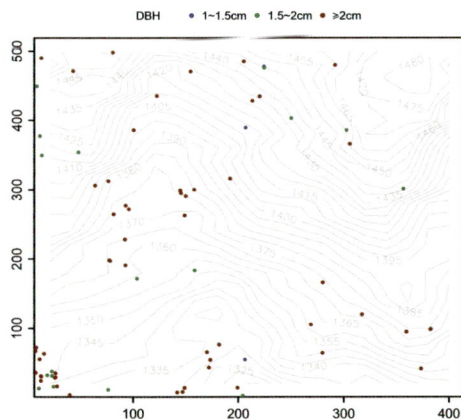

个体分布图 Distribution of individuals

径级分布表 DBH class

胸径区间 (Diameter class) (cm)	个体数 (No. of individuals in the plot)	比例 (Proportion) (%)
1～2	20	34.48
2～5	34	58.62
5～10	3	5.17
10～20	1	1.73
20～30	0	0.00
30～60	0	0.00
≥60	0	0.00

45 北京忍冬

Lonicera elisae Franch.
忍冬科 | Caprifoliaceae

代码 (SpCode) = LONELI

个体数 (Individual number/20 hm^2) = 4

最大胸径 (Max DBH) = 2.9 cm

重要值排序 (Importance value rank) = 47

落叶灌木。叶对生，叶片卵状椭圆形，两面生柔毛，下面带灰色。双花并生；相邻两花的萼筒分离；花冠漏斗状，白色或淡粉色，外面有毛，基部具浅囊。浆果椭球形，成熟时红色，离生。花期4～5月，果期5～6月。

Deciduous shrubs. Leaves are opposite, the leaves are ovate oval, soft hairs on both sides, and gray underneath. Double flowers coexist; separation of calyxes adjacent to two flowers; corolla funnel-shaped, white or pale pink, hairy on the outside, with shallow sacs at the base. Berries are ellipsoidal, red when ripe, and detached. Fl. Apr.-May, fr. May-Jun..

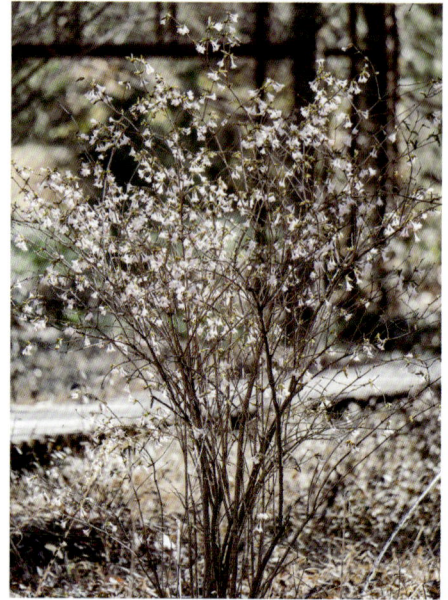

植株　　　　　Whole plant
摄影：林秦文　　Photo by: Lin Qinwen

叶与果序　　　Leaves and infructescences
摄影：林秦文　　Photo by: Lin Qinwen

花序　　　　　Inflorescences
摄影：林秦文　　Photo by: Lin Qinwen

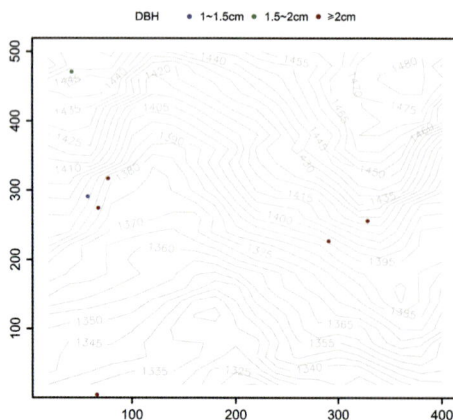

个体分布图 Distribution of individuals

径级分布表 DBH class

胸径区间 (Diameter class) (cm)	个体数 (No. of individuals in the plot)	比例 (Proportion) (%)
1～2	1	25.00
2～5	3	75.00
5～10	0	0.00
10～20	0	0.00
20～30	0	0.00
30～60	0	0.00
≥60	0	0.00

46 鸡树条荚蒾

Viburnum opulus subsp. *calvescens* (Rehder) Sugim.
忍冬科 | Caprifoliaceae

代码 (SpCode) = VIBOPU

个体数 (Individual number/20 hm^2) = 5

最大胸径 (Max DBH) = 3.4 cm

重要值排序 (Importance value rank) = 49

落叶灌木，达1.5~4 m。树冠圆球形。树皮暗灰色，常纵裂。小枝有棱，皮孔显著。冬芽卵圆形。叶对生，圆卵形，常3裂，边缘具不整齐粗牙齿，两面无毛，具掌状3出脉；花药紫红色。核果近圆形，红色。花期5~6月，果期9~10月。

Deciduous shrubs, up to 1.5-4 m tall. Crown spherical. Bark is dark gray, often longitudinally lobed. Twigs are ridged and the skin holes are prominent. Winter buds are oval. Leaves are opposite, round and ovate, often 3 lobes, the edges are irregular and coarse teeth, the sides are hairless, and have palmate 3 veins; anthers are purple red. The stone fruits are nearly round and red. Fl. May-Jun., fr. Sep.-Oct..

植株　　　　Whole plant
摄影：林秦文　　Photo by: Lin Qinwen

花序　　　　Inflorescences
摄影：林秦文　　Photo by: Lin Qinwen

果序　　　　Infructescences
摄影：林秦文　　Photo by: Lin Qinwen

径级分布表 DBH class

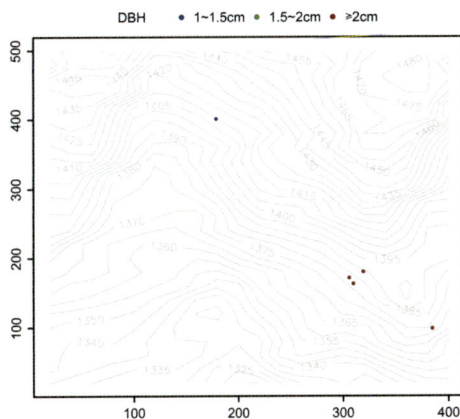

个体分布图 Distribution of individuals

胸径区间 (Diameter class) (cm)	个体数 (No. of individuals in the plot)	比例 (Proportion) (%)
1~2	1	20.00
2~5	4	80.00
5~10	0	0.00
10~20	0	0.00
20~30	0	0.00
30~60	0	0.00
≥60	0	0.00

47　接骨木

Sambucus williamsii Hance
忍冬科 | Caprifoliaceae

代码 (SpCode) = SAMWIL
个体数 (Individual number/20 hm^2) = 8
最大胸径 (Max DBH) = 8.6 cm
重要值排序 (Importance value rank) = 42

落叶灌木至小乔木，达6 m。树冠宽卵形。树皮暗灰色，纵裂。小枝粗壮，皮孔显著，髓心淡黄棕色。单数羽状复叶对生，小叶常5~7，椭圆形至矩圆状披针形，无毛，边缘有锯齿，揉碎后有臭味。圆锥花序顶生。花白色至淡黄色。浆果状核果近球形，黑紫色或红色。花期5~6月，果期7~8月。

Deciduous shrubs to small trees, up to 6 m tall. Crown broadly ovate. Bark is dark gray, longitudinally lobed. Twigs are thick, the skin holes are prominent, and the marrow heart is yellowish brown. Singular pinnate compound leaves are paraphyletic, the leaflets are often 5-7, oval to rectangular lanceolate, glabrous, serrated at the edges, and smelly after crushing. Conical inflorescence apical. Flowers are white to pale yellow. Berry-like drupes are nearly spherical, black-purple or red. Fl. May-Jun., fr. Jul.-Aug..

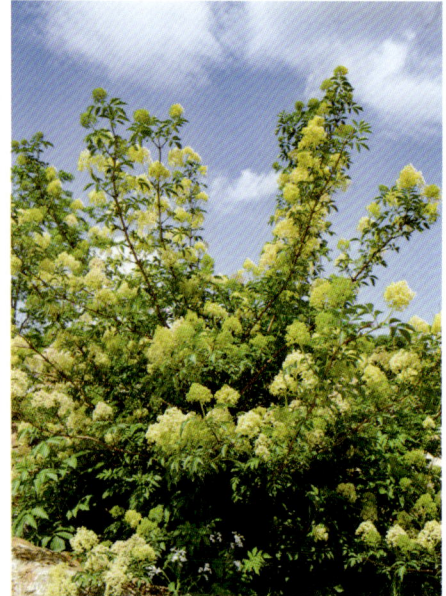

植株　　Whole plant
摄影：林秦文　　Photo by: Lin Qinwen

叶与花序　　Leaves and inflorescences
摄影：林秦文　　Photo by: Lin Qinwen

果序　　Infructescences
摄影：林秦文　　Photo by: Lin Qinwen

径级分布表　DBH class

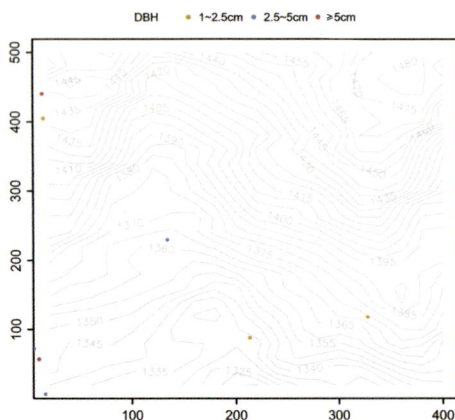

个体分布图 Distribution of individuals

胸径区间 (Diameter class) (cm)	个体数 (No. of individuals in the plot)	比例 (Proportion) (%)
1~2	1	12.50
2~5	6	75.00
5~10	1	12.50
10~20	0	0.00
20~30	0	0.00
30~60	0	0.00
≥60	0	0.00

48 刺五加

Eleutherococcus senticosus (Rupr. et Maxim.) Maxim.
五加科 | Araliaceae

代码 (SpCode) = ELESEN

个体数 (Individual number/20 hm^2) = 32

最大胸径 (Max DBH) = 3.4 cm

重要值排序 (Importance value rank) = 39

落叶灌木，2～4 m。树冠宽卵形。树皮灰白色，具细密刺。小枝常被密刺。掌状复叶互生，小叶常5，有时3，纸质，椭圆状倒卵形至矩圆形，边缘具锐尖重锯齿。伞形花序单个顶生或2～4个聚生，具多花；花瓣5，卵形，淡绿色。浆果状核果几球形至卵形。花期6～7月，果期8～10月。

Deciduous shrubs, 2-4 m tall. Crown broadly ovate. Bark is greyish white with fine spines. Twigs are often densely pricked. Palmate compound leaves are alternate, leaflets are often 5, sometimes 3, papery, oval inverted ovate to rectangular, with sharp sharp and heavy serrations at the edges. Umbel-shaped inflorescences are single apical or 2-4 polyphyletic, with multiple flowers; petal 5, ovate, pale green. Berry-like drupes are spherical to ovate. Fl. Jun.-Jul., fr. Aug.-Oct..

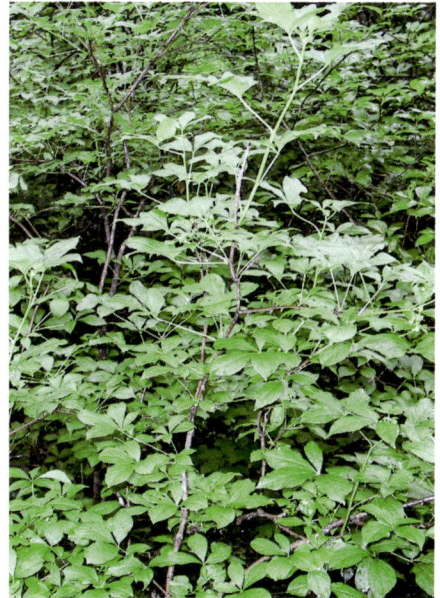

植株　　　　　　　Whole plant
摄影：林秦文　　　Photo by: Lin Qinwen

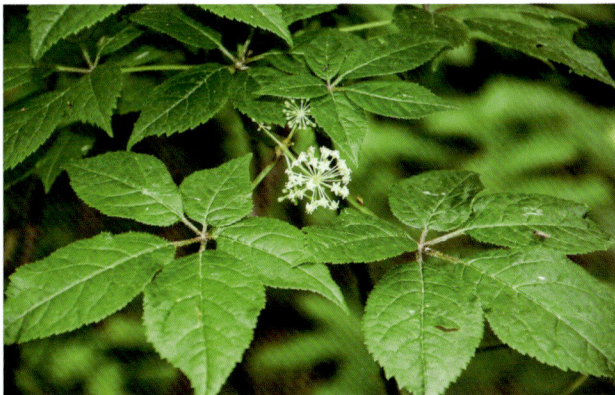

叶与花序　　　　　　Leaves and inflorescences
摄影：叶建飞　　　　Photo by: Ye Jianfei

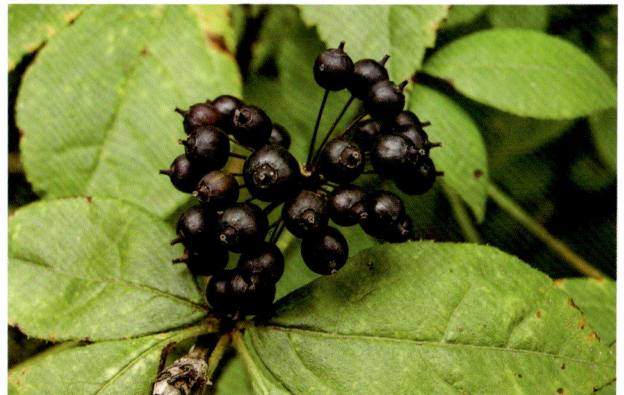

果序　　　　　　　Infructescences
摄影：刘冰　　　　Photo by: Liu Bing

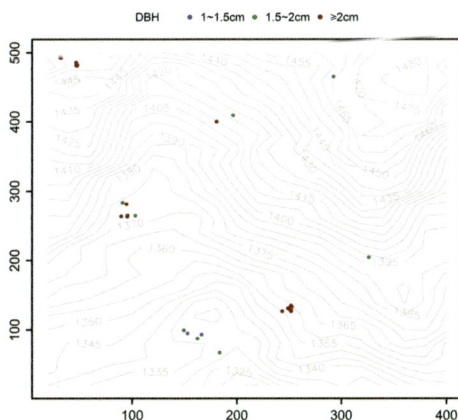

个体分布图 Distribution of individuals

径级分布表 DBH class

胸径区间 (Diameter class) (cm)	个体数 (No. of individuals in the plot)	比例 (Proportion) (%)
1～2	17	53.12
2～5	15	46.88
5～10	0	0.00
10～20	0	0.00
20～30	0	0.00
30～60	0	0.00
≥60	0	0.00

东灵山样地：草本植物及其分布
Donglingshan Plot:
Herbaceous Plants and Their Distribution

4

49 荚果蕨 jiá guǒ jué | Ostrich Fern

Matteuccia struthiopteris (L.) Tod.
球子蕨科 | Onocleaceae

最大高度 (Max height) = 50 cm
重要值排序 (Importance value rank) = 27

多年生草本，植株高70～110 cm。根状茎直立。营养叶倒披针形或长椭圆形，2回羽状深裂。能育叶较短，有粗硬而较长的柄，羽状，倒披针形，羽片深棕色，线状念珠状，硬化，很多内卷，隐藏在孢子堆。

Perennial herbs, 70-110 cm tall. Rhizome erect. Sterile leaves oblanceolate or oblong, 2-pinnate-pinnatifid. Fertile lamina shorter than sterile lamina and have stiff stalks, pinnate, oblanceolate, dark brown, oblanceolate, dark brown, linear-moniliform, hardened, much inrolled, hidden in spore piles.

孢子叶　Fertile lamina
摄影：林秦文　Photo by: Lin Qinwen

植株　Whole plant
摄影：刘博　Photo by: Liu Bo

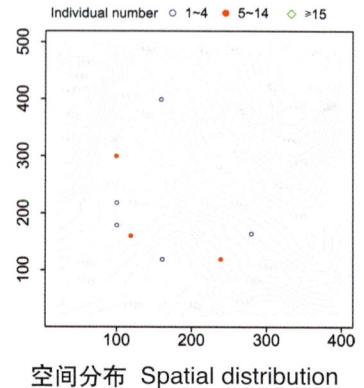

空间分布　Spatial distribution

50 蕨 jué | China Brackenfern

Pteridium aquilinum var. *latiusculum* (Desv.) Underw. ex A. Heller
碗蕨科 | Dennstaedtiaceae

最大高度 (Max height) = 74 cm
重要值排序 (Importance value rank) = 25

多年生草本。叶直立，0.5～1.5 cm。叶片3或4羽状裂，三角形到长圆形卵形的轮廓，革质；裂片存在于2个相邻节段之间；羽片上升或水平，卵状三角形到长圆形，高达40×15 cm，先端锐尖；羽片或裂片线形到长圆形，除羽片边缘和中脉外无毛。

Perennial herbs. Fronds erect, 0.5-1.5 cm tall. Lamina 3 or 4 pinnate-pinnatifid, triangular to oblong-ovate in outline, leathery; lobes present between 2 adjacent segments; pinnae ascending or horizontal, ovate-triangular to oblong, up to 40 × 15 cm, apex acute; pinnules or segments linear to oblong, glabrous except for pinnule margins and midvein.

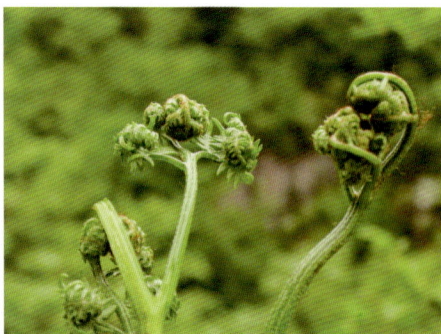

拳卷叶　Young fronds
摄影：林秦文　Photo by: Lin Qinwen

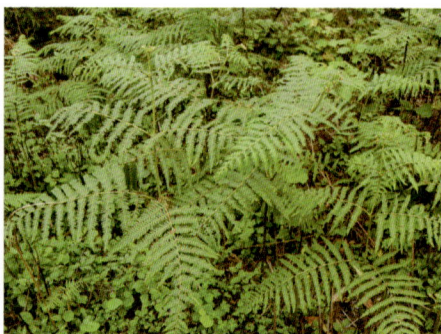

植株　Whole plant
摄影：林秦文　Photo by: Lin Qinwen

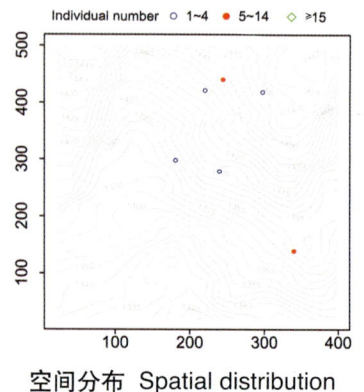

空间分布　Spatial distribution

51 草问荆

Equisetum pratense Ehrh.
木贼科 | Equisetaceae

最大高度 (Max height) = 32 cm
重要值排序 (Importance value rank) = 64

多年生草本。可育茎禾草色，高15～25 cm，直径2～2.5 mm。根状茎直立或匍匐，黑棕色，节和根具稀疏的长黄棕色毛状体或无毛。球果椭圆形圆柱状，1～2.2 cm，直径3～7 mm，先端钝；成熟时茎延长，1.7～4.5 cm。

Perennial herbs. Fertile stems straw-colored, 15-25 cm tall, 2-2.5 mm in diam.. Rhizome erect or creeping, blackish brown, nodes and roots with sparse long yellowish brown trichomes or glabrous. Strobilus ellipsoid-terete, 1-2.2 cm, 3-7 mm in diam., apex blunt; stalk prolonged when mature and 1.7-4.5 cm.

果序 Infructescences
摄影：林秦文 Photo by: Lin Qinwen

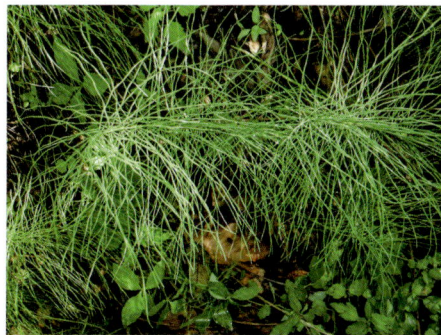

植株 Whole plant
摄影：林秦文 Photo by: Lin Qinwen

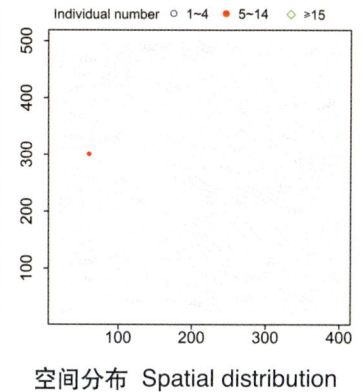

空间分布 Spatial distribution

52 穿龙薯蓣

Dioscorea nipponica Makino
薯蓣科 | Dioscoreaceae

最大高度 (Max height) = 150 cm
重要值排序 (Importance value rank) = 13

多年生草本，长达500 cm。茎向左旋而缠绕。单叶互生，叶片广卵形或卵形，基部心形。花雌雄异株，雄花序复穗状，有花2～7朵；雌花序穗状，下垂花小钟形。蒴果倒卵状椭圆形。种子具膜质翅。花期5～7月，果期7～10月。

Perennial herbs, up to 500 cm long. Stem twining to left. Leaves alternate, simple, leaves shiny, base cordate. Male flowers, usually in cymules or umbellules of 2-7; female flowers spike, nodding, flower campanulate. Capsule ellipsoid-oblanceolate. Seeds winged all round. Fl. May-Jul., fr. Jul.-Oct..

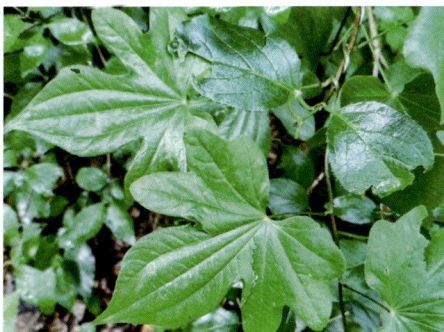

叶 Leaves
摄影：刘博 Photo by: Liu Bo

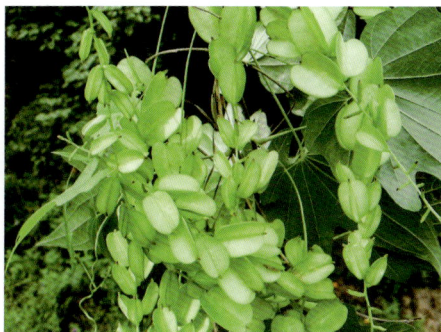

果序 Infructescences
摄影：林秦文 Photo by: Lin Qinwen

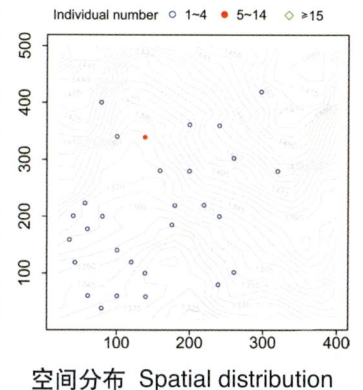

空间分布 Spatial distribution

53 藜芦 lí lú | Black False Hellebore

Veratrum nigrum L.
藜芦科 | Melanthiaceae

多年生草本，高达100 cm。茎直立。叶互生，无柄或者在茎末梢有很短的柄。圆锥花序顶生，侧枝一般为雄花，圆锥顶端一般为两性花。花被片紫黑色，长圆形。蒴果卵球形。花期7～8月，果期8～9月。

最大高度 (Max height) = 35 cm
重要值排序 (Importance value rank) = 20

Perennial herbs, to 100 cm tall. Stem erect. Leaves alternate, sessile or with a short stalk at the end of the stem. Panicles terminal, lateral branches are generally male flowers, and the top is generally bisexual flowers. Tepals black-purple, oblong. Capsules ovoid. Fl. Jul.-Aug., fr. Aug.-Sep..

花序　Inflorescences
摄影：林秦文　Photo by: Lin Qinwen

植株　Whole plant
摄影：林秦文　Photo by: Lin Qinwen

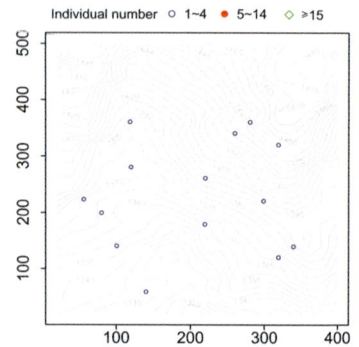

Individual number ○ 1~4 ● 5~14 ◇ ≥15

空间分布　Spatial distribution

54 北重楼 běi chóng lóu | Verticillate

Paris verticillata M. Bieb.
藜芦科 | Melanthiaceae

多年生草本，高25～60 cm。叶柄非常短；叶轮生，狭长圆形，倒披针形或倒卵形，基部楔形。花序梗4.5～12 cm，外部花被通常4（或5），绿色，长圆状披针形或卵形披针形，内部黄绿色。浆果紫色、黑色，球状，直径约1 cm。种子无假种皮。花期5～6月，果期7～9月。

最大高度 (Max height) = 22 cm
重要值排序 (Importance value rank) = 87

Perennial herbs, 25-60 cm tall. Petioles very short; whorl leaf, narrowly oblong, oblanceolate, or obovate-oblanceolate, base cuneate. Outer tepals usually 4 (or 5), green, oblong-lanceolate or ovate-lanceolate, inner ones yellow-green. Berry purple-black, globose, ca. 1 cm in diam.. Seeds without aril. Fl. May-Jun., fr. Jul.-Sep..

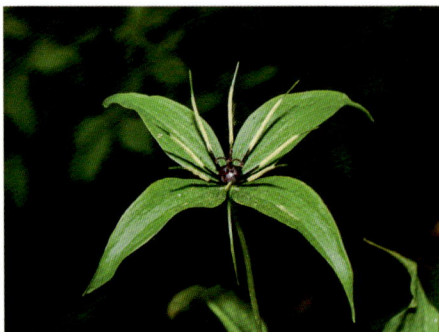

花　Flower
摄影：林秦文　Photo by: Lin Qinwen

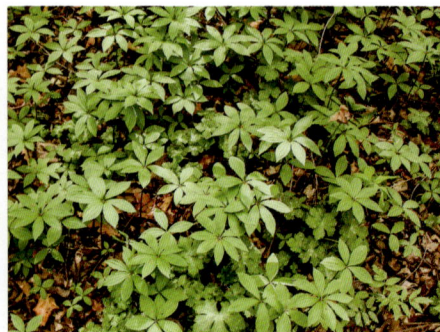

植株　Whole plant
摄影：林秦文　Photo by: Lin Qinwen

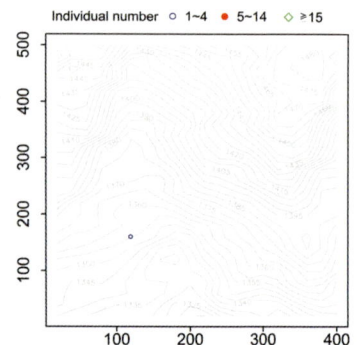

Individual number ○ 1~4 ● 5~14 ◇ ≥15

空间分布　Spatial distribution

55 野鸢尾

yě yuān wěi | Vesper Iris

Iris dichotoma Pall.
鸢尾科 | Iridaceae

多年生草本。根状茎不规则块状，棕褐色。叶基部扇形，在下部花茎上互生，灰绿色，剑形，稍弯曲。花茎二歧分枝。花紫罗兰色，淡蓝色或奶油色具紫棕色斑点。花被裂片为6，有少数条纹。蒴果圆柱形，长3.5～5 cm。种子椭圆形，暗褐色，有小翅。花期7～8月，果期8～9月。

最大高度 (Max height) = 32 cm
重要值排序 (Importance value rank) = 51

Perennial herbs. Rhizome irregular massive, brown. Leaves in basal fans and alternate on flowering stems proximally, grayish green, sword-shaped, slightly curved. Flowering stems dichotomously branched. Flowers violet, pale blue, or cream with purplish brown markings. Perianth lobes 6, with a few stripes. Capsule cylindrical, 3.5-5cm long. Seeds elliptic, dark brown, with small wings. Fl. Jul.-Aug., fr. Aug.-Sep..

花序　Inflorescences
摄影：林秦文　Photo by: Lin Qinwen

植株　Whole plant
摄影：林秦文　Photo by: Lin Qinwen

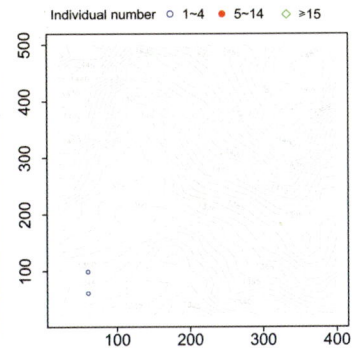

空间分布　Spatial distribution

56 茖葱

gè cōng | Alpine Leek

Allium victorialis L.
石蒜科 | Amaryllidaceae

多年生草本，高4～6 cm。具根状茎，鳞茎单生或数枚聚生。叶2～3枚，叶片倒披针状椭圆形至椭圆形。伞形花序球形，花白色或带绿色；花被片6。蒴果倒心形，种子卵形，黑色。花期5月，果期8月。

最大高度 (Max height) = 35 cm
重要值排序 (Importance value rank) = 16

Perennial herbs, 4-6 cm tall. Bulb solitary or clustered. Leaves 2-3, oblanceolate-elliptic to elliptic. Umbel spherical, with white or greenish flowers; perianth 6. Capsule obcordate, seed ovoid, black. Fl. May, fr. Aug..

花序　Inflorescence
摄影：林秦文　Photo by: Lin Qinwen

植株　Whole plant
摄影：林秦文　Photo by: Lin Qinwen

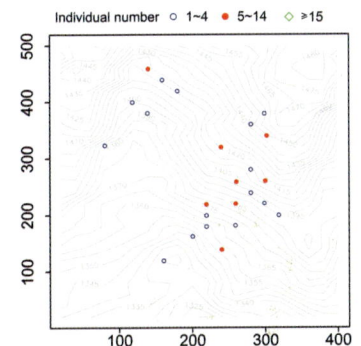

空间分布　Spatial distribution

57 龙须菜

Asparagus schoberioides Kunth.
天门冬科 | Asparagaceae

最大高度 (Max height) = 54 cm
重要值排序 (Importance value rank) = 23

多年生草本，高达100 cm。茎直立，无刺，上部和分枝具纵棱，分枝有时有极窄的翅。鳞叶近披针形。花序在分枝后发育，腋生。两性花2～4簇生，近等长。浆果径约6 mm，成熟时红色，具1～2种子。花期5～6月，果期8～9月。

Perennial herbs, 100 cm tall. Stems erect, unarmed, distinctly striate-ridged distally, branches angled or sometimes narrowly winged. Scaly leaves sublanceolate. Inflorescences developing after cladodes, axillary. Flowers of both sexes in clusters of 2-4, subequal. Berries ca. 6 mm in diam., red when ripe, with 1-2 seeds. Fl. May-Jun., fr. Aug.-Sep..

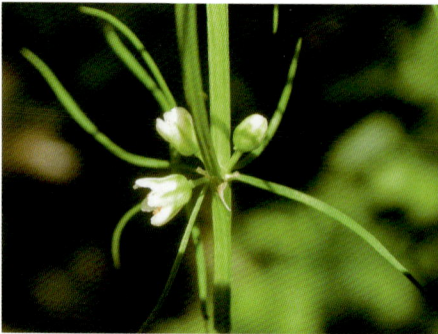

花序　Inflorescences
摄影：林秦文　Photo by: Lin Qinwen

果序　Infructescences
摄影：林秦文　Photo by: Lin Qinwen

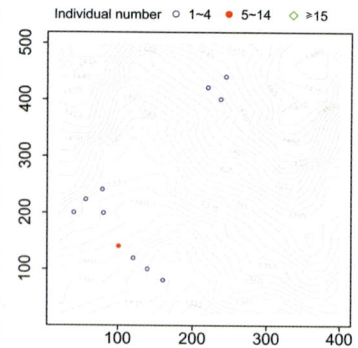

空间分布　Spatial distribution

58 铃兰

Convallaria keiskei Miq.
天门冬科 | Asparagaceae

最大高度 (Max height) = 30 cm
重要值排序 (Importance value rank) = 26

多年生草本，高18～30 cm，光滑。叶片背面光滑，椭圆形至卵状披针形，基部楔形，顶部削尖。花茎轻微弯曲，先端近尖到渐尖。总状花序，花朵钟状，下垂，6～10朵，乳白色。浆果灰白色。花期5～6月，果期7～9月。

Perennial herbs, 18-30 cm tall, glabrous. Leaves abaxially glaucescent, elliptic to ovate-lanceolate, base cuneate, Pedicel slightly curved, apex subacute to acuminate. Racemes, 6-10 flowed, campanulate, nodding. Perianth white. Berry off white. Fl. May-Jun., fr. Jul.-Sep..

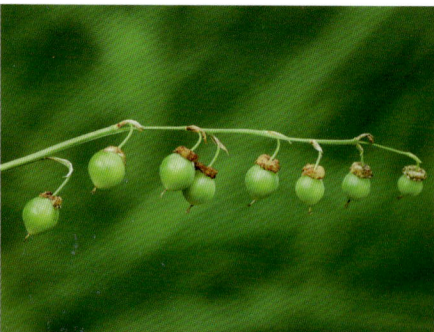

果序　Infructescences
摄影：林秦文　Photo by: Lin Qinwen

花序　Inflorescences
摄影：林秦文　Photo by: Lin Qinwen

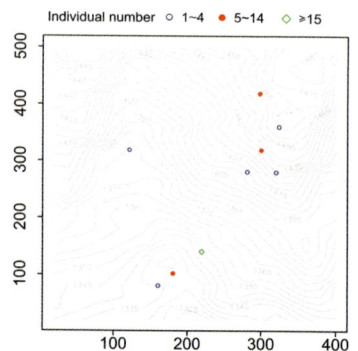

空间分布　Spatial distribution

59　鹿药

Maianthemum japonicum (A. Gray) LaFrankie
天门冬科 | Asparagaceae

多年生草本，高20～40 cm。茎单生，密生粗毛。叶互生，通常4～7枚。花10～20朵排成圆锥花序，花白色，被片离生或稍合生在基部，长圆形或倒卵形。浆果近球形，红色，具1～2颗种子，球形。花期5～6月，果期8月。

最大高度 (Max height) = 36 cm
重要值排序 (Importance value rank) = 33

Perennial herbs, 20-40 cm tall. Stem solitary, pubescent. Leaves alternate, usually 4-7. Inflorescence paniculate, 10-20 flowered, flowers white, segments free or slightly connate at base, oblong or oblong-obovate. Fruits globose, red at maturity, seeds 1-2, compressed globose. Fl. May-Jun., fr. Aug..

花序　Inflorescences
摄影：林秦文　Photo by: Lin Qinwen

植株　Whole plant
摄影：林秦文　Photo by: Lin Qinwen

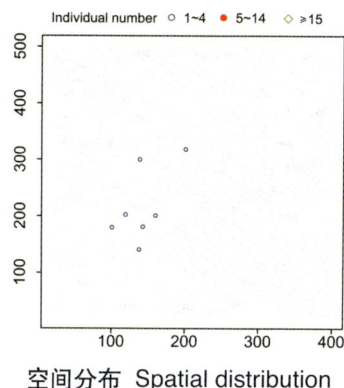

空间分布　Spatial distribution

60　舞鹤草

Maianthemum bifolium (L.) F. W. Schmidt
百合科 | Liliaceae

多年生草本，高8～20 cm，无毛。基生叶1枚，在花期枯萎。茎生叶通常2枚，互生于茎的上部。花通常10～25朵排成总状花序，花白色，花被片4，长圆形。浆果红色，种子外种皮黄色。花期5～6月，果期8～9月。

最大高度 (Max height) = 6 cm
重要值排序 (Importance value rank) = 43

Perennial herbs, 8-20 cm tall, glabrous. Basal leaf 1, withered at anthesis. Cauline leaves usually 2, borne distally to apically on stem. Raceme erect, 10-25-flowered, flowers white, perianth 4, oblong. Berries red at maturity. Seeds with yellow testa. Fl. May-Jun., fr. Aug.-Sep..

果序　Infructescences
摄影：林秦文　Photo by: Lin Qinwen

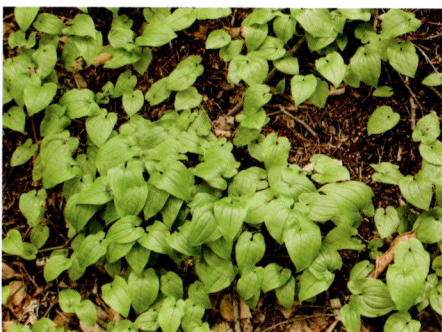

植株　Whole plant
摄影：林秦文　Photo by: Lin Qinwen

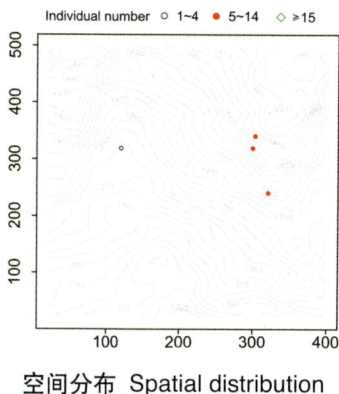

空间分布　Spatial distribution

77

61 玉竹

Polygonatum odoratum (Mill.) Druce
天门冬科 | Asparagaceae

最大高度 (Max height) = 78 cm
重要值排序 (Importance value rank) = 9

多年生草本，高20～70 cm。茎单一，弯曲，无毛，具有棱角。叶通常为7～12枚，互生，叶片椭圆形、近披针形或卵形。花序通常具1～4朵花，花下垂；花被一般6片，白色或淡黄绿色，圆柱状或钟状圆形。浆果蓝黑色，直径7～10 mm。花期5～6月，果期7～9月。

Perennial herbs, 20-70 cm tall. Stem arching, glabrous, angled. Leaves 7-12, alternate. Leaves elliptic to ovate-oblong. Inflorescences 1-4 flowered, flowers pendulous; perianth 6, white to yellowish green, cylindric or campanulate. Berries blue-black, 7-10 mm in diam.. Fl. May-Jun., fr. Jul.-Sep..

花序　Inflorescences
摄影：林秦文　Photo by: Lin Qinwen

果序　Infructescences
摄影：林秦文　Photo by: Lin Qinwen

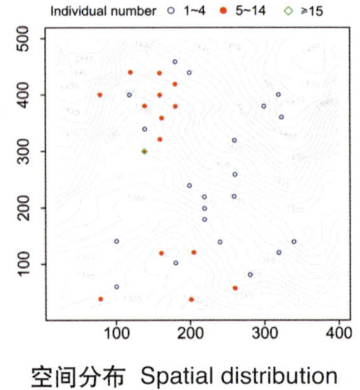

空间分布　Spatial distribution

62 黄精

Polygonatum sibiricum Delar. ex Redoute
天门冬科 | Asparagaceae

最大高度 (Max height) = 35 cm
重要值排序 (Importance value rank) = 89

多年生草本。根状茎圆柱状，茎高50～90 cm，有时呈攀援状。叶轮生，每轮4～6枚，无柄，条状披针形，先端拳卷或弯曲成钩。花序通常具2～4朵花，似成伞形，总花梗长1～2 cm，苞片位于花梗基部，钻形或条状披针形。花被乳白色至淡黄色，花被筒中部稍缢缩。浆果径7～10 mm，成熟时黑色，种子4～7。花期5～6月，果期8～9月。

Perennial herbs. Rhizomes terete, 50-90 cm tall, sometimes climbing. Leaves in whorls of 4-6, sessile, linear-lanceolate, glabrous, apex strongly cirrose or curved. Inflorescences umbel-like, usually 2-4-flowered, peduncle 1-2 cm, bracts borne at base of pedicel, subulate to linear-lanceolate. Perianth milky white to pale yellow, cylindric, slightly constricted in middle. Berries 7-10 mm in diam., black, with 4-7 seeds. Fl. May-Jun., fr. Aug.-Sep..

花序　Inflorescences
摄影：林秦文　Photo by: Lin Qinwen

果序　Infructescences
摄影：刘博　Photo by: Liu Bo

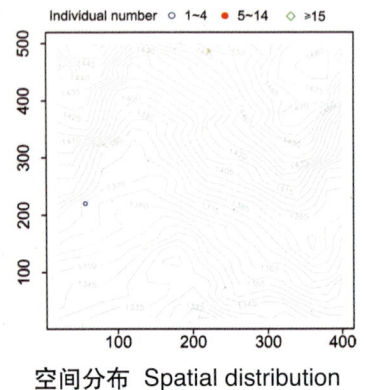

空间分布　Spatial distribution

63　细叶薹草

Carex duriuscula subsp. *stenophylloides* (V. I. Krecz.) S. Yun Liang　莎草科 | Cyperaceae

最大高度 (Max height) = 42 cm
重要值排序 (Importance value rank) = 4

多年生草本。叶片扁平。花序带褐色。本亚种与原亚种之区别在于果囊较大，长3.5～4.5 mm，卵形或卵状椭圆形，雌性颖片棕色栗色，通常短于胞果，很少等长，具狭窄的透明边缘，顶端渐狭成较长的喙。花果期4～6月。

Perennial herbs. Leaves flat. Inflorescence brownish. The difference between this subspecies and the original subspecies is that the fruit sac is larger, 3.5-4.5 mm long, oval or oval in shape, female glumes brown-castaneous, usually shorter than utricles, rarely as long, with narrower hyaline margins, the tip gradually narrows into a longer beak. Fl. and fr. Apr.-Jun..

果序　Infructescences
摄影：姚红霞　Photo by: Yao Hongxia

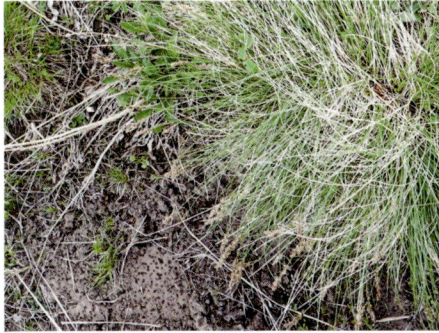

植株　Whole plant
摄影：姚红霞　Photo by: Yao Hongxia

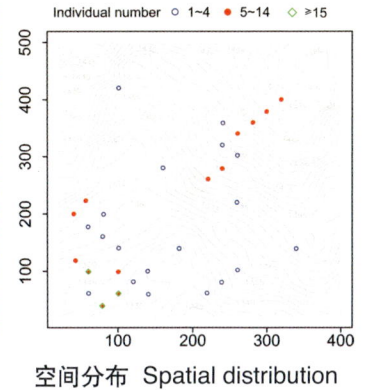

空间分布　Spatial distribution

64　宽叶薹草

Carex siderosticta Hance
莎草科 | Cyperaceae

最大高度 (Max height) = 25 cm
重要值排序 (Importance value rank) = 32

多年生草本。花茎和营养茎间隔；花茎被无叶片鞘覆盖在基部，浅棕色，没有叶。花茎可达30 cm高；穗状花序3～6，单个或二联在每个节上，雌雄同体，或顶生穗状花序通常雄性，线形圆筒状，1.5～3 cm，松散开花。紧密包裹的小坚果，椭圆形，三棱，约2 mm。花果期4～5月。

Perennial herbs. Flowering culms and vegetative culms spaced; flowering culms clothed by bladeless sheaths at base, pale brown, without leaves. Flowering culms up to 30 cm tall; spikes 3-6, single or binate at each node, androgynous, or terminal spike usually male, linear-cylindric, 1.5-3 cm, loosely flowercd. Nutlet tightly envelopcd, elliptic, trigonous, ca. 2 mm. Fl. and fr. Apr.-May.

果序　Inflorescences
摄影：林秦文　Photo by: Lin Qinwen

植株　Whole plant
摄影：林秦文　Photo by: Lin Qinwen

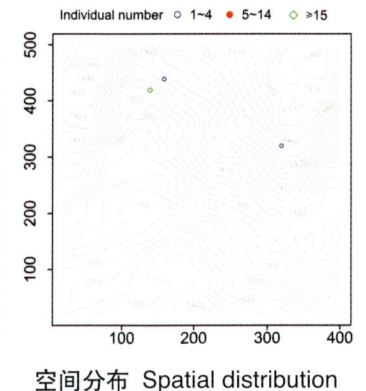

空间分布　Spatial distribution

65 大披针薹草

Carex lanceolata Boott
莎草科 | Cyperaceae

多年生草本，高10～35 cm。叶宽1～2.5 mm，扁平、柔软，边缘较为粗糙。顶生雄性穗状花序，线形圆筒状；侧穗状花序2～5，雌性，长圆形或圆筒状长圆形。雌性颖片侧向紫棕色，纸质，3脉，边缘宽白色透明先端锐尖。小坚果倒卵状椭圆形，三棱，基部具短柄，先端具下弯和短喙。花果期4～7月。

最大高度 (Max height) = 35 cm
重要值排序 (Importance value rank) = 10

Perennial herbs, 10-35 cm tall. Leaves 1-2.5 mm wide, flat, soft, margins slightly scabrous. Terminal spike male, linear-cylindric; lateral spikes 2-5, female, oblong or oblong-cylindric. Female glumes purple-brown laterally, 3-veined, margins broadly white hyaline, apex acute, mucronate. Nutlets obovate-elliptic, trigonous, base shortly stipitate, apex with recurved and short beak. Fl. and fr. Apr.-Jul..

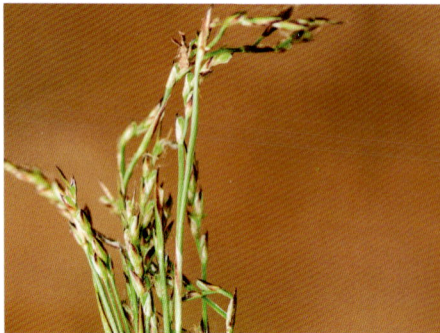

果序　Infructescences
摄影：林秦文　Photo by: Lin Qinwen

植株　Whole plant
摄影：林秦文　Photo by: Lin Qinwen

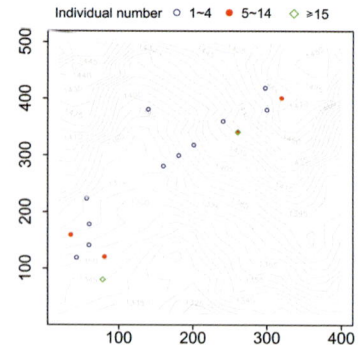

Individual number ○ 1~4　● 5~14　◇ ≥15

空间分布　Spatial distribution

66 野青茅

Calamagrostis arundinacea (L.) Roth
禾本科 | Poaceae

多年生草本。簇生，有时具短根状茎。秆直立，纤细或粗壮，高(40～)100～150 cm，节2～3。叶鞘无毛到密被短柔毛；叶片平或内卷，平滑，或短柔毛。圆锥花序轮廓卵状披针形。小穗3～5(～6.5) mm，黄绿色或紫绿色。花果期7～10月。

最大高度 (Max height) = 50 cm
重要值排序 (Importance value rank) = 1

Perennial herbs. Tufted, sometimes shortly rhizomatous. Culms erect, slender or robust, (40-) 100-150 cm tall, 2-3-noded. Leaf sheaths glabrous to densely pubescent; leaves flat or involute, smooth, scabrid or pubescent. Panicle outline ovate-lanceolate. Spikelets 3-5 (-6.5) mm, yellowish green or purplish green. Fl. and fr. Jul.-Oct..

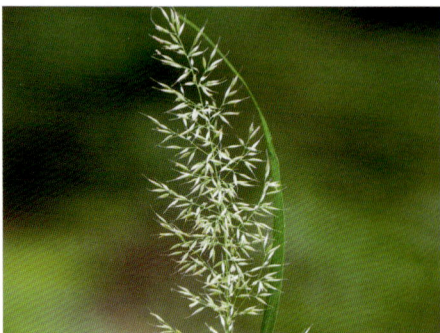

花序　Inflorescences
摄影：林秦文　Photo by: Lin Qinwen

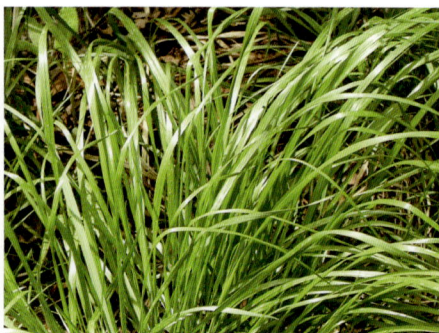

植株　Whole plant
摄影：林秦文　Photo by: Lin Qinwen

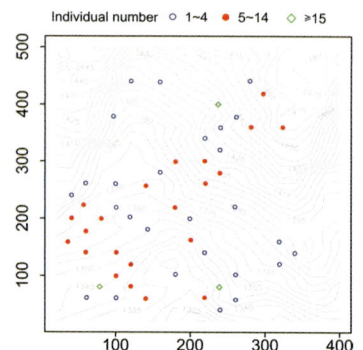

Individual number ○ 1~4　● 5~14　◇ ≥15

空间分布　Spatial distribution

67　大油芒

dà yóu máng | Frost Grass

Spodiopogon sibiricus Trin.
禾本科 | Poaceae

最大高度 (Max height) = 78 cm
重要值排序 (Importance value rank) = 15

多年生草本，高70～150 cm。根状茎被鳞。叶片线状披针形，无毛或短柔毛，先端刚毛渐尖。圆锥花序疏散收缩，轮廓狭披针形。总状花序2～3节，具7～9小穗，成熟时脱节，一对小穗，一个无梗，另一个则有花梗。宽披针形的下部颖片，先端锐尖或稍微缺。花果期7～10月。

Perennial herbs, 70-150 cm tall. Spreading scaly rhizomes. Leaves linear-lanceolate, glabrous or pubescent, apex setaceously acuminate. Panicle loosely contracted, narrowly lanceolate in outline. Racemes 2-3-noded with 7-9 spikelets, disarticulating at maturity, one spikelet of a pair sessile, the other pedicellate. lower glume broadly lanceolate, apex acute or slightly emarginate. Fl. and fr. Jul.-Oct..

果序　Infructescences
摄影：林秦文　Photo by: Lin Qinwen

植株　Whole plant
摄影：林秦文　Photo by: Lin Qinwen

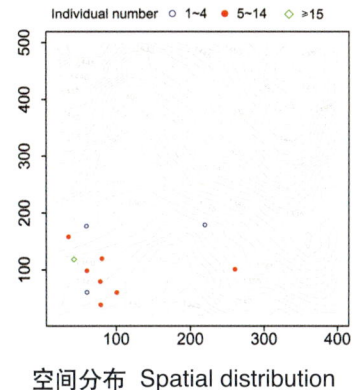

空间分布　Spatial distribution

68　小黄紫堇

xiǎo huáng zǐ jǐn | Radde's Fumewort

Corydalis raddeana Regel
罂粟科 | Papaveraceae

最大高度 (Max height) = 15 cm
重要值排序 (Importance value rank) = 60

多年生草本，高90 cm。茎直立。基生叶2～3回羽状分裂，具长柄，三角形或宽卵形。茎生叶多数，下部者具长柄，上部者具短柄，其他与基生叶相同。总状花序顶生或腋生。种子近圆形，直径1.5～2 mm，黑色，具光泽。花果期6～10月。

Perennial herbs, 90 cm tall. Stems erect. Basal leaves 2-3-ternate-pinnate, stipitat, triangular or broadly ovate. Cauline leaves majority, lower part with long petiole, upper part with short petiole, other same as basal leaves. Racemes terminal or axillary. Seeds suborbicular, 1.5-2 mm in diam., black, lustrous. Fl. and fr. Jun.-Oct..

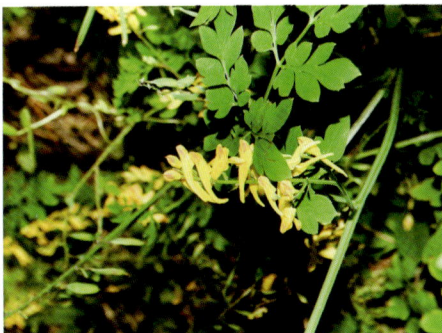

花序　Inflorescences
摄影：林秦文　Photo by: Lin Qinwen

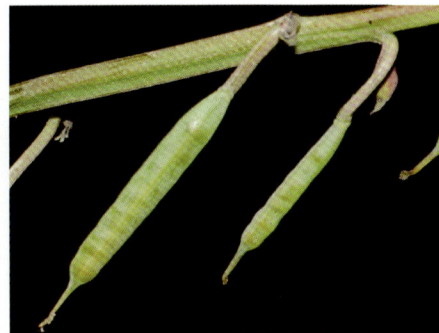

果序　Infructescences
摄影：林秦文　Photo by: Lin Qinwen

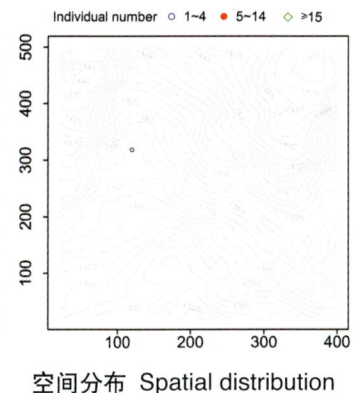

空间分布　Spatial distribution

69 东亚唐松草

dōng yà táng sōng cǎo | White-backed Small Meadow-rue

Thalictrum minus var. *hypoleucum* (Siebold & Zucc.) Miq.
毛茛科 | Ranunculaceae

最大高度 (Max height) = 120 cm
重要值排序 (Importance value rank) = 11

多年生草本，高达150 cm。叶为3至4回三出羽状复叶，脉网明显。圆锥状花序，具多花或少花；萼片4，绿白色，早落，卵形，长3～4 mm。雄蕊多数，花药狭长圆形，花丝上部倒披针形。瘦果卵球形，长2～3 mm。花果期6～7月。

Perennial herbs, 150 cm tall. Leaves 3 or 4 ternately pinnate, reticulate veins. Panicle, with many or few flowers; sepals 4, greenish white, caducous, ovate, 3 to 4 mm long. Stamens numerous, filament oblanceolate, anther narrowly oblong. Achenes ovoid, 2-3 mm long. Fl. and fr. Jun.-Jul..

叶　　　Leaves
摄影：刘博　　　Photo by: Liu Bo

花序　　　Inflorescence
摄影：林秦文　　　Photo by: Lin Qinwen

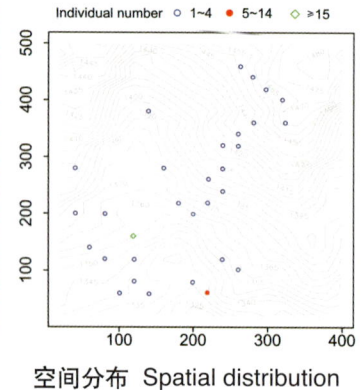

Individual number ○ 1~4　● 5~14　◇ ≥15

空间分布　Spatial distribution

70 华北耧斗菜

huá běi lóu dòu cài | Easten Red Columbine

Aquilegia yabeana Kitag.
毛茛科 | Ranunculaceae

最大高度 (Max height) = 20 cm
重要值排序 (Importance value rank) = 28

多年生草本，茎高40～60 cm。常在上部分枝，除被柔毛外还密被腺毛。基生叶少数，2回三出复叶，正面无毛，背面被短柔毛，叶柄长达8～25 cm。花序少数花，苞片三裂或不裂，萼片紫色，狭卵形，花瓣直立，倒卵形，比萼片稍长或稍短，顶端近截形，距直或微弯。种子黑色，狭卵球形。花果期5～6月。

Perennial herbs, stems 40-60 cm tall. Pubescent or densely glandular hairy, often apically branched. Basal leaves few, 2-ternate, petiole to 8-25 cm, leaves abaxially pubescent to subglabrous, adaxially glabrous. Inflorescences few flowered, bracts trilobed or unlobed, sepals purple, narrowly ovate, petals erect, obovate, nearly as long as sepals, apex subtruncate; straight or apically slightly incurved. Seeds black, narrowly ovoid. Fl. and fr. May-Jun..

花序　　　Inflorescences
摄影：林秦文　　　Photo by: Lin Qinwen

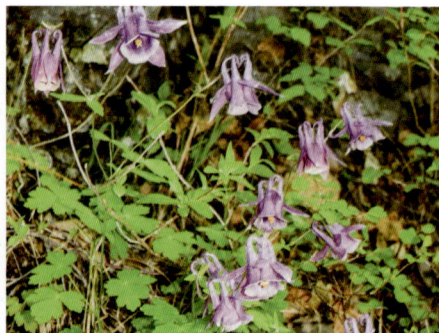

植株　　　Whole plant
摄影：林秦文　　　Photo by: Lin Qinwen

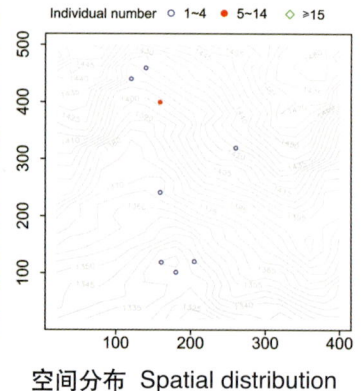

Individual number ○ 1~4　● 5~14　◇ ≥15

空间分布　Spatial distribution

71 两色乌头

liǎng sè wū tóu | Bicolor Monkshood

Aconitum alboviolaceum Kom.
毛茛科 | Ranunculaceae

一年生或多年生草本。根圆柱状，长200～400 cm。茎缠绕，疏被短柔毛或无毛。叶片五角状肾形，三深裂稍超过中部或近中部。花3～8朵，萼片淡紫色或近白色，被伸展的柔毛。种子大约2.5 mm。花期8～9月。

最大高度 (Max height) = 70 cm
重要值排序 (Importance value rank) = 35

Annual or perennial herbs. Rhizome terete, 200-400 cm long. Stems twining, sparsely pubescent or glabrous. The leaves is pentagonal, reniform, and has three deep lobes, slightly exceeding or near the middle. Inflorescences 3-8-flowered, sepals lavender or nearly white, extended pilose. Follicles erect, ca. 2.5 mm. Fl. Aug.-Sep..

花序　Inflorescences
摄影：林秦文　Photo by: Lin Qinwen

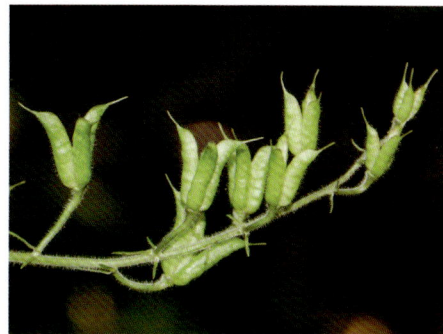

果序　Infructescences
摄影：林秦文　Photo by: Lin Qinwen

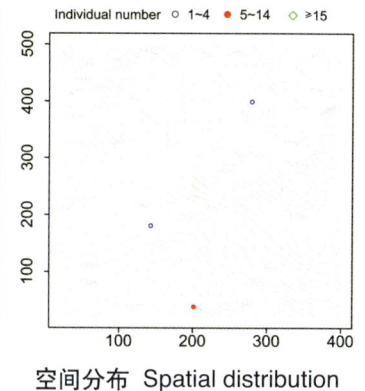

空间分布　Spatial distribution

72 北乌头

běi wū tóu | Kusnezoff's Monkshood

Aconitum kusnezoffii Rchb.
毛茛科 | Ranunculaceae

多年生草本，高80～150 cm。叶柄3～11 cm，无毛；叶片五边形，纸质的或近革质，背面无毛，正面疏生反折短柔毛，基部心形。花序顶生，9～22花。下部的花梗1.8～3.5 cm，具2小苞片在中部或下部。蓇葖果直立，种子约2.5 mm。花期7～9月。

最大高度 (Max height) = 17 cm
重要值排序 (Importance value rank) = 79

Perennial herbs, 80-150 cm tall. Petiole 3-11 cm, glabrous; leaves pentagonal, papery or subleathery, abaxially glabrous, adaxially sparsely retrorse pubescent, base cordate. Inflorescence terminal, 9-22 flowered. Proximal pedicels 1.8-3.5 cm, with 2 bracteoles at middle or below. Follicles erect, seeds ca. 2.5 mm. Fl. Jul.-Sep..

花序　Inflorescences
摄影：林秦文　Photo by: Lin Qinwen

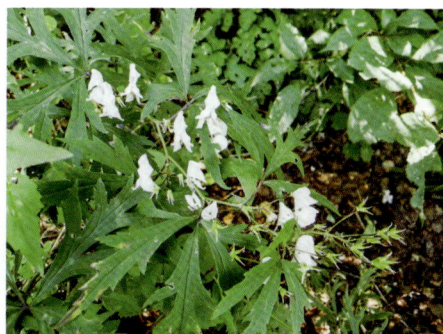

植株　Whole plant
摄影：林秦文　Photo by: Lin Qinwen

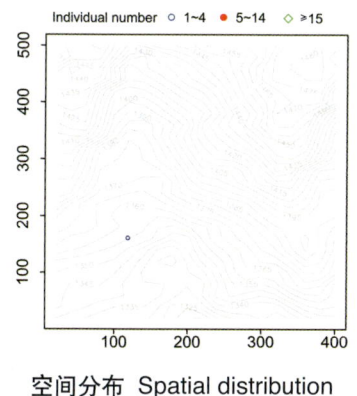

空间分布　Spatial distribution

73　高乌头

Aconitum sinomontanum Nakai
毛茛科 | Ranunculaceae

最大高度 (Max height) = 10 cm

重要值排序 (Importance value rank) = 73

多年生草本。根状茎圆柱状约20 cm，以及下部的茎生叶具长叶柄；叶柄30～50 cm，近无毛；叶片肾形或圆形，两面稀疏具短柔毛或脱落无毛，基部心形，3深裂；狭楔形菱形的中央裂片，3半裂具不规则的三角形的齿在边缘，先端渐尖。菁葵果1.1～1.7 cm。种子倒卵形，约3 mm。花期6～9月，果期9月。

Perennial herbs. Rhizome terete, ca. 20 cm, and proximal cauline leaves long petiolate; petiole 30-50 cm, nearly glabrous; leaf blade reniform or orbicular-reniform, both surfaces sparsely pubescent or glabrescent, base cordate, 3-parted; central lobe narrowly cuneate-rhombic, 3-fid with irregular triangular teeth at margin, apex acuminate. Follicles 1.1-1.7 cm. Seeds obovate, ca. 3 mm. Fl. Jun.-Sep., fr. Sep..

花序　　Inflorescences
摄影：林秦文　Photo by: Lin Qinwen

植株　　Whole plant
摄影：林秦文　Photo by: Lin Qinwen

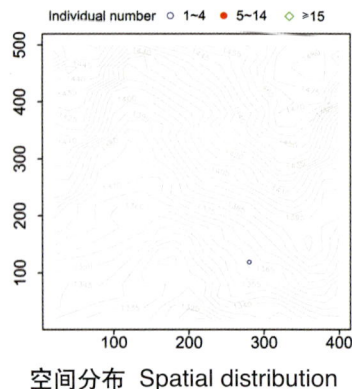

Individual number　○ 1~4　● 5~14　◇ ≥15

空间分布　Spatial distribution

74　牛扁

Aconitum barbatum var. *puberulum* Ledeb.
毛茛科 | Ranunculaceae

最大高度 (Max height) = 91 cm

重要值排序 (Importance value rank) = 19

多年生草本，茎高55～90 cm。茎和叶柄均被反曲而紧贴的短柔毛。叶片圆肾形，中全裂片分裂不近中脉，末回小裂片三角形或狭披针形；2回裂片具狭卵型小裂片，叶面具有绿白色斑点。总状花序，轴及花梗密被紧贴的短柔毛，花梗直展；萼片黄色，外面密被短柔毛；花瓣无毛，距与花瓣长近等长。花期8～9月，果期9～10月。

Perennial herbs, stem 55-90 cm tall. Stem and petioles retrorse and appressed pubescent. Central leaf segment 3-parted not nearly to midvein, ultimate lobes triangular or narrowly lanceolate; 2-lobes are narrow egg shaped small lobes with green and white spots on the leaf surface. Racemes, rachis and pedicels densely appressed pubescent, pedicels spreading straight. Upper sepal cylindric. The petals are glabrous and the flower pitch is nearly equal to the petal length. Fl. Aug.-Sep., fr. Sep.-Oct..

果序　　Infructescences
摄影：林秦文　Photo by: Lin Qinwen

植株　　Whole plant
摄影：林秦文　Photo by: Lin Qinwen

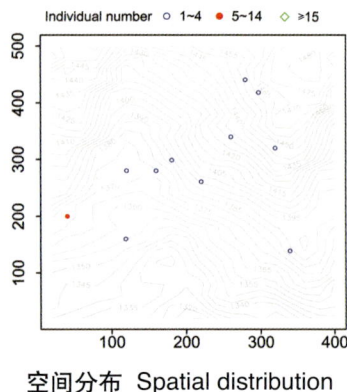

Individual number　○ 1~4　● 5~14　◇ ≥15

空间分布　Spatial distribution

75 类叶升麻

lèi yè shēng má | Chinese Bugbane

Actaea asiatica H.Hara
毛茛科 | Ranunculaceae

最大高度 (Max height) = 29 cm
重要值排序 (Importance value rank) = 57

多年生草本，高30～80 cm。圆柱状，4～6 mm直径，不分枝，基部无毛，顶部白色短柔毛。叶2或3，具长叶柄；叶柄10～17 cm。花瓣匙形，2～2.5 mm；花丝3～5 mm；花药长约0.7 mm。果单生，紫色黑色。花期5～6月，果期7～9月。

Perennial herbs, 30-80 cm tall. Terete, 4-6 mm in diam., unbranched, basally glabrous, apically white pubescent. Leaves 2 or 3, long petiolate; petiole 10-17 cm. Petals spatulate, 2-2.5 mm; filaments 3-5 mm; anthers ca. 0.7 mm. Fruit solitary, purple-black. Fl. May-Jun., fr. Jul.-Sep..

果序　Infructescences
摄影：姚红霞　Photo by: Yao Hongxia

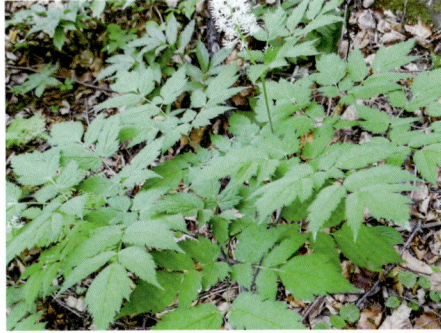

花序　Inflorescences
摄影：林秦文　Photo by: Lin Qinwen

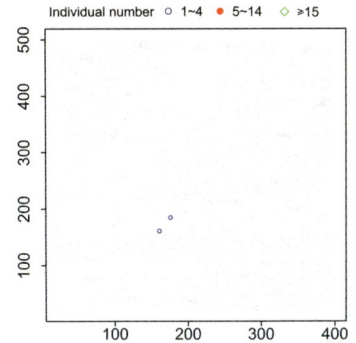

空间分布　Spatial distribution

76 棉团铁线莲

mián tuán tiě xiàn lián | Six-petal

Clematis hexapetala Pall.
毛茛科 | Ranunculaceae

最大高度 (Max height) = 70 cm
重要值排序 (Importance value rank) = 54

多年生草本，高达100 cm。茎疏被柔毛。叶为1至2回羽状全裂，线状披针形、线形或椭圆形。聚伞花序顶生或腋生，1至3多花；苞片叶状或披针形；花梗长1～7 cm；萼片白色，长椭圆形或狭倒卵形；花蕾时期呈棉球状，雄蕊无毛。瘦果倒卵圆形，长2.5～3.5 mm。花期6～8月，果期7～10月。

Perennial herbs, 100 cm tall. Stems sparsely pubescent. Leaves 1-2-pinnate-pinnatifid, linear-lanceolate, linear or ellipse. Cymes terminal or axillary, 1-3 to many flowered; bracts foliaceous or lanceolate; peduncle 1-7 cm; sepals white, spreading, narrowly obovate to narrowly oblong. The flower buds are like cotton balls, and the stamens are glabrous. Achenes obovate, 2.5-3.5 mm. Fl. Jun.-Aug., fr. Jul. -Oct..

花序　Inflorescences
摄影：林秦文　Photo by: Lin Qinwen

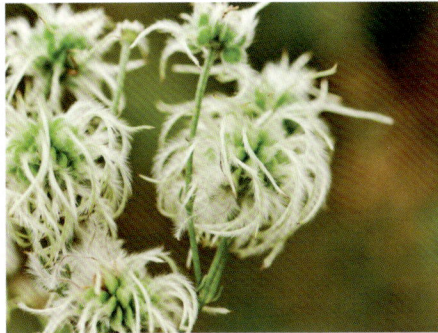

果序　Infructescences
摄影：刘博　Photo by: Liu Bo

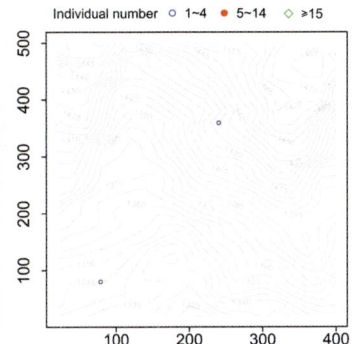

空间分布　Spatial distribution

77　大叶铁线莲

dà yè tiě xiàn lián | Hyacinth-flower Clematis

Clematis heracleifolia DC.
毛茛科 | Ranunculaceae

最大高度 (Max height) = 22 cm
重要值排序 (Importance value rank) = 80

亚灌木或者多年生草本，高30~100 cm。茎通常分枝。叶三出；叶柄2.5~14 cm，被微柔毛；小叶叶片宽卵形，五边形，椭圆形，或卵形，纸质，通常3浅裂，后脱落。复聚伞花序顶生或腋生；花序梗4~8 cm，密被微柔毛；花梗0.8~3.5 cm，密被微柔毛到茸毛。花柱3~4 mm，密被长柔毛。瘦果椭圆，被微柔毛。花期8~9月，果期10月。

Subshrubs or perennial herbs, 30-100 cm tall. Stems usually branched. Leaves ternate; petiole 2.5-14 cm, puberulous; leaflet blades broadly ovate, pentagonal, suborbicular, elliptic, or ovate, papery, often 3 lobed, glabrescent. Compound cymes terminal or axillary, peduncle 4-8 cm, densely puberulous. Pedicel 0.8-3.5 cm, densely puberulous to velutinous. Style 3-4 mm, densely villous. Achenes elliptic, puberulous. Fl. Aug.-Sep., fr. Oct..

花序　Inflorescences
摄影：林秦文　Photo by: Lin Qinwen

植株　Whole plant
摄影：林秦文　Photo by: Lin Qinwen

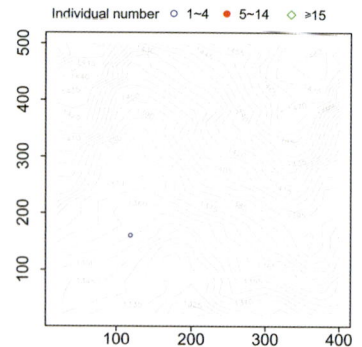

空间分布　Spatial distribution

78　毛金腰

máo jīn yāo | Dense-pilose Golden-sexifrage

Chrysosplenium pilosum Maxim.
虎耳草科 | Saxifragaceae

最大高度 (Max height) = 4 cm
重要值排序 (Importance value rank) = 61

多年生草本，高14~16 cm。茎棕色柔毛。茎生叶相反。叶柄3.5 mm，棕色柔毛；叶片扇形，两面和边缘无毛或棕色具柔毛，基部楔形，波状的边缘不明显6或清楚钝牙齿，先端近截形。蒴果约5.5 mm。种子暗褐色，宽椭圆形，约1 mm，约17具槽，脊具小乳突。聚伞花序长约2厘米；花梗无毛。花柱长约1 mm。花果期4~6月。

Perennial herbs, 14-16 cm tall. Stems brown pilose. Cauline leaves opposite. Petiole ca. 3.5 mm, brown pilose; leaves flabellate, both surfaces and margin glabrous or brown pilose, base cuneate, margin obscurely 6-undulate-crenate or distinctly obtusely dentate, apex subtruncate. Capsule ca. 5.5 mm. Seeds dark brown, broadly ellipsoid, ca. 1 mm, ca. 17-sulcate, ridges papillose. Cymes 2 cm long, peduncle glabrous. Style ca. 1mm. Fl. and fr. Apr.-Jun..

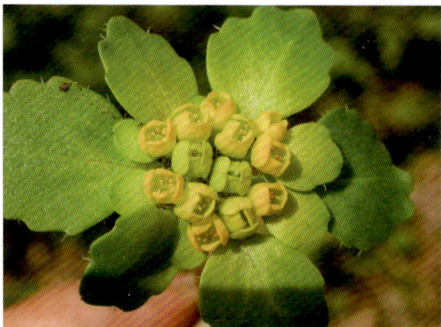

花序　Inflorescences
摄影：林秦文　Photo by: Lin Qinwen

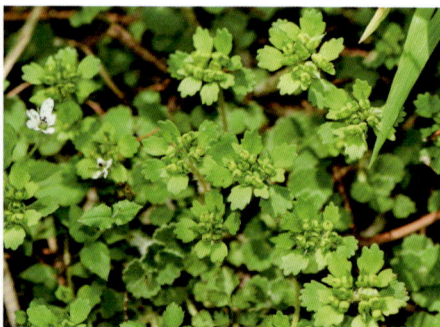

植株　Whole plant
摄影：林秦文　Photo by: Lin Qinwen

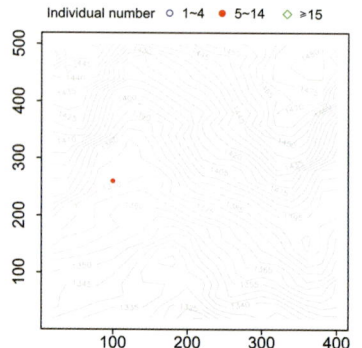

空间分布　Spatial distribution

79 落新妇

Astilbe chinensis (Maxim.) Franch. et Sav.
虎耳草科 | Saxifragaceae

最大高度 (Max height) = 42 cm
重要值排序 (Importance value rank) = 55

多年生草本，高50~100 cm。茎直立，无毛。基生叶2~3回三出复叶，小叶卵状长圆形、菱状卵形或卵形。茎生叶2~3，小叶膜质，褐色。顶生圆锥花序，较狭，密被褐色细长的卷曲柔毛；苞片卵形；花瓣5，紫色，线形。蒴果长约3 mm。种子褐色。花期7~8月，果期9月。

Perennial herbs, 50-100 cm tall. Stems erect, glabrous. Basal leaves 2 or 3 ternately compound, leaflets rhombic elliptic or ovate to elliptic. Cauline leaves 2 or 3. Densely flowered; densely brown long crisped hairy; bracts ovate, petals 5, lilac to purple, linear. Capsule ca. 3 mm. Seeds brown. Fl. Jul.-Aug., fr. Sep..

花序 Inflorescences
摄影：林秦文 Photo by: Lin Qinwen

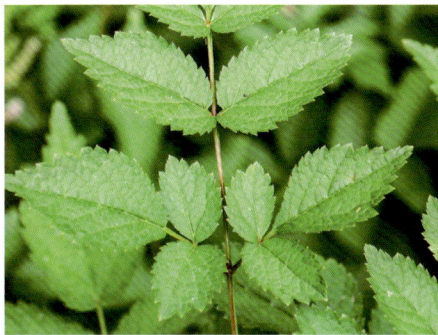

叶 Leaves
摄影：林秦文 Photo by: Lin Qinwen

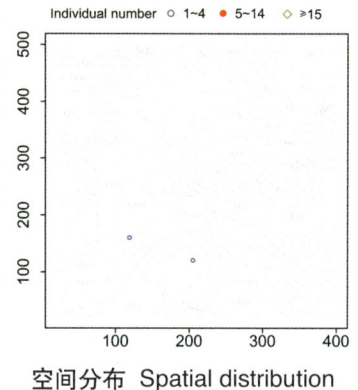

空间分布 Spatial distribution

80 费菜

Phedimus aizoon (L.) 't Hart
景天科 | Crassulaceae

最大高度 (Max height) = 11 cm
重要值排序 (Importance value rank) = 47

多年生草本。块根胡萝卜状。茎高达50 cm，无毛，不分枝。叶互生，狭披针形、椭圆状披针形至卵状倒披针形，先端渐尖，基部楔形，边缘有不整齐的锯齿。聚伞花序有多花，水平分枝。花瓣5，黄色，长圆形至椭圆状披针形。种子椭圆形。花期6~7月，果期8~9月。

Perennial herbs. Roots tuberous, carrot-shaped. Stems, simple, erect, 50 cm tall. Leaves alternate, leaves narrowly lanceolate, elliptic-lanceolate, elliptic, ovate-oblanceolate, base cuneate, margin irregularly serrate, apex obtuse-rounded or acuminate. Inflorescence horizontally branched, many flowered. Flowers unequally 5-merous, petals yellow, oblong to elliptic-lanceolate, apex mucronate. Seeds elliptic. Fl. Jun.-Jul., fr. Aug.-Sep..

花序 Inflorescences
摄影：林秦文 Photo by: Lin Qinwen

植株 Whole plant
摄影：林秦文 Photo by: Lin Qinwen

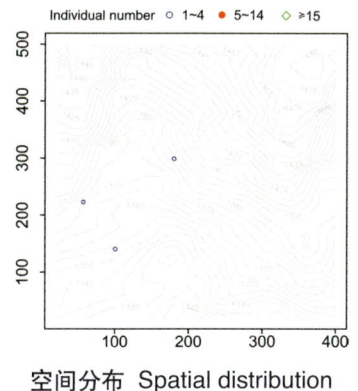

空间分布 Spatial distribution

81　草珠黄芪　　　　　　　　　　cǎo zhū huáng qí | Grass beads astragalus

Astragalus capillipes Fisch. ex Bunge
豆科 | Fabaceae

多年生草本，高40～50 cm。茎直立，35～40 cm。叶2～5 cm，无毛。总状花序3～14（～20）cm，松散多花；花序梗2～9 cm，近无毛或疏生短茸毛；花瓣白色，奶油色，橙色，或红色；标准近圆形，约6 mm×4 mm，先端微缺。花期7～9月，果期9～10月。

最大高度 (Max height) = 51 cm
重要值排序 (Importance value rank) = 69

Perennial herbs, 40-50 cm tall. Stems erect, 35-40 cm. Leaves 2-5 cm, glabrous. Racemes 3-14 (-20) cm, loosely many flowered; peduncle 2-9 cm, subglabrous or sparsely short tomentose; petals white, cream, orange, or red; standard suborbicular, ca. 6 mm × 4 mm, apex emarginate. Fl. Jul.-Sep., fr. Sep.-Oct..

花序　Inflorescences
摄影：林秦文　Photo by: Lin Qinwen

植株　Whole plant
摄影：姚红霞　Photo by: Yao Hongxia

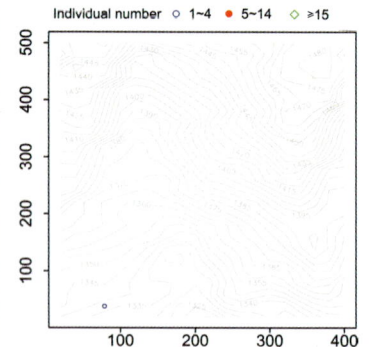

空间分布　Spatial distribution

82　山野豌豆　　　　　　　　　　shān yě wān dòu | Pleasant Vetch

Vicia amoena Fisch. ex Ser.
豆科 | Fabaceae

多年生草本，高30～100 cm。植株被疏柔毛，稀近无毛。主根粗壮，须根发达。偶数羽状复叶近无柄，小叶4～7对，椭圆形至卵披针形。总状花序通常长于叶。花冠红紫色、蓝紫色或蓝色花期颜色多变。荚果长圆形，两端渐尖，无毛。种子圆形，种皮革质，深褐色，具花斑。花期4～6月，果期7～10月。

最大高度 (Max height) = 55 cm
重要值排序 (Importance value rank) = 68

Perennial herbs, 30-100 cm tall. Pilose, rarely densely sericeous. Taproot stout, fibrous roots developed. Stem much branched, slender. Leaves subsessile, paripinnate, leaflets 4-7-paired, elliptic to ovate-lanceolate, apex rounded and retuse. Raceme usually longer than leaf. Corolla blue, blue-purple, red-purple. Pod oblong, both ends acu-minate, glabrous. Seeds round, kind of leathery, dark brown, with flower spots. Fl. Apr.-Jun., fr. Jul.-Oct..

花序　Inflorescences
摄影：林秦文　Photo by: Lin Qinwen

植株　Whole plant
摄影：林秦文　Photo by: Lin Qinwen

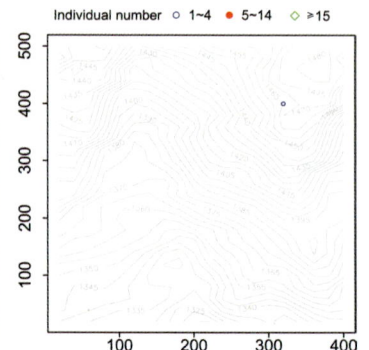

空间分布　Spatial distribution

83 歪头菜

wāi tóu cài | Two-leaf Vetch

Vicia unijuga A.Braun
豆科 | Fabaceae

多年生草本，高40～100 cm。根茎近木质。叶偶数羽状，托叶戟形或近披针形，边缘小齿状。小叶一对，卵状披针形或近菱形，先端渐尖，边缘具小齿状，基部楔形，两面均疏被微柔毛。总状花序单一，稀有分支；花萼紫色，斜钟状或钟状；花冠蓝紫色，较少为白色。荚果扁、长圆形，无毛。花期6～7月，果期8～9月。

最大高度 (Max height) = 32 cm
重要值排序 (Importance value rank) = 46

Perennial herbs, 40-100 cm tall. Stem sub-shrubby. Leaves paripinnate, stipules hastate or sublanceolate, margin unequally toothed. Leaflets 1-paired, ovate to lanceolate or rhombic-ellip-tic, both surfaces pilose, apex sometimes acuminate. Raceme rarely branched. Calyx campanulate or obliquely so, glabrescent. Pod flat, oblong, glabrous. Fl. Jun.-Jul., fr. Aug.-Sep..

花序 Inflorescences
摄影：林秦文 Photo by: Lin Qinwen

果序 Infructescences
摄影：林秦文 Photo by: Lin Qinwen

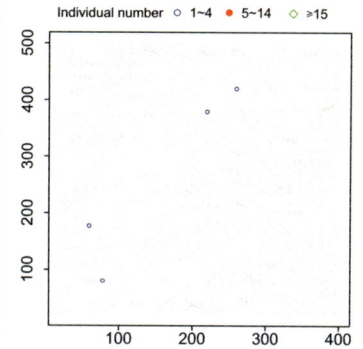

Individual number ○ 1~4 ● 5~14 ◇ ≥15

空间分布 Spatial distribution

84 远志

yuǎn zhì | Thin-leaf Milkwort

Polygala tenuifolia Willd.
远志科 | Polygalaceae

多年生草本，高15～50 cm。茎多分枝，丛生，直立或倾斜，具脊和槽，短柔毛。叶近无柄，叶片线形到线状披针形，纸质、无毛或具非常稀疏柔毛，背面中脉突出，正面凹陷，侧脉不明显，基部楔形，边缘全缘，下弯，先端渐尖。总状花序顶生。蒴果圆形，具狭翅，无纤毛，先端微凹。种子黑色，卵球形。花果期5～9月。

最大高度 (Max height) = 30 cm
重要值排序 (Importance value rank) = 71

Perennial herbs, 15-50 cm tall. Stems much branched, caespitose, erect or inclined, ridged and sulcate, pubescent. Leaves subsessile, leaves linear to linear-lanceolate, papery, glabrous or very sparsely puberulent, midvein raised abaxially, impressed adaxially, lateral veins obscure, base cuneate, margin entire, recurved, apex acuminate. Raccmes terminal. Capsule orbicular, narrowly winged, eciliate, apex retuse. Seeds black, ovoid. Fl. and fr. May-Sep..

花序 Inflorescences
摄影：林秦文 Photo by: Lin Qinwen

植株 Whole plant
摄影：林秦文 Photo by: Lin Qinwen

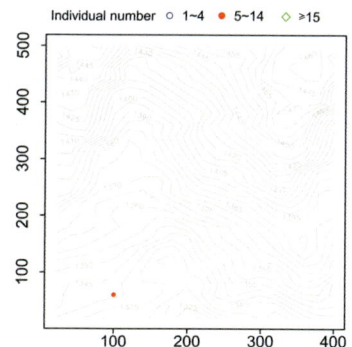

Individual number ○ 1~4 ● 5~14 ◇ ≥15

空间分布 Spatial distribution

85 龙牙草

Agrimonia pilosa Ledeb.
蔷薇科 | Rosaceae

最大高度 (Max height) = 84 cm
重要值排序 (Importance value rank) = 50

多年生草本，高30～120 cm。根状茎短，通常块茎状，叶柄疏生柔毛或短柔毛；叶片间断奇数羽状具3或4小叶，退化至3小叶于上部叶片。花直径6～9 mm；花梗1～5 mm，具柔毛；苞片通常3与的裂片分别线形。花瓣黄色，长圆形。果托杯倒卵球形的圆锥形。花果期5～12月。

Perennial herbs, 30-120 cm tall. Rhizome short, usually tuberous, leaves interrupted imparipinnate with 3 or 4 pairs of leaflets, reduced to 3 leaflets on upper leaves. Flowers 6-9 mm in diam; pedicel 1-5 mm, pilose; bract usually 3 parted with segments linear. Petals yellow, oblong, stigma capitate. Fruiting hypanthium obovoid-conic. Fl. and fr. May-Dec..

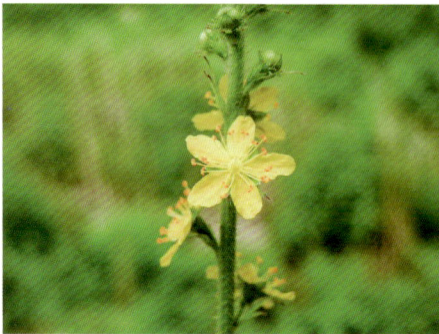

花序 Inflorescences
摄影：林秦文 Photo by: Lin Qinwen

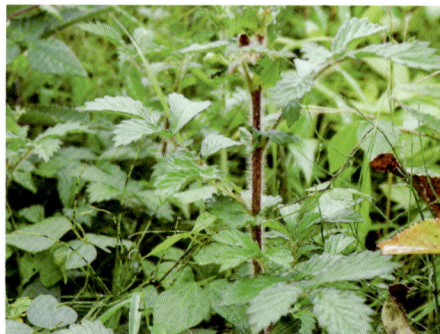

植株 Whole plant
摄影：林秦文 Photo by: Lin Qinwen

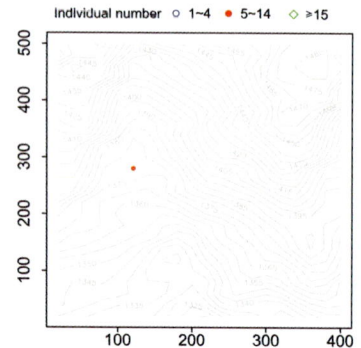

空间分布 Spatial distribution

86 地榆

Sanguisorba officinalis L.
蔷薇科 | Rosaceae

最大高度 (Max height) = 35 cm
重要值排序 (Importance value rank) = 30

多年生草本，高12 cm，有棱。基生叶为羽状复叶，叶柄长，无毛或疏生腺体，基部具鞘和覆瓦状，有时疏生腺毛；叶片具4～6对小叶；小叶具小叶柄，两面为绿色，卵形，或扁化披针形。穗状花序椭圆形或圆柱形；萼片4，紫红色，椭圆形或宽卵形，背面被疏柔毛。瘦果有4棱。花果期7～10月。

Perennial herbs, 12 cm tall, stems angulate. Basal leaves pinnate compound leaves, petiole long, glabrous or sparsely glandular, base sheathing and imbricate, sometimes sparsely glandular hairy; leaves with 4-6 pairs of leaflets; leaflets petiolulate, green on both surfaces, ovate, or fasciated lanceolate. Inflorescences elliptic, cylindrical or ovate; sepals 4, purplish red, elliptic or broadly ovate, abaxially sparsely pilose. Achene, 4-ribbed. Fl. and fr. Jul.-Oct..

花序 Inflorescences
摄影：林秦文 Photo by: Lin Qinwen

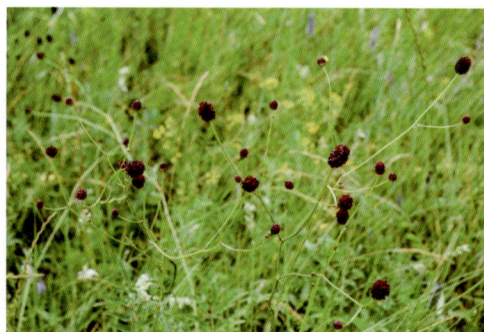

植株 Whole plant
摄影：林秦文 Photo by: Lin Qinwen

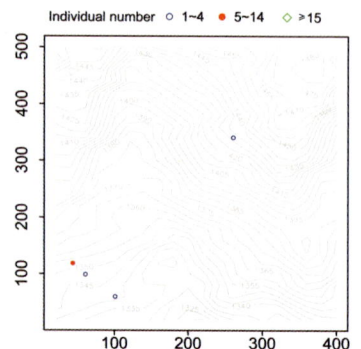

空间分布 Spatial distribution

87 菊叶委陵菜

Potentilla tanacetifolia Willd. ex D.F.K.Schltdl.
蔷薇科 | Rosaceae

多年生草本。根粗壮，圆柱形。花茎直立或上升，高15~65 cm。基生羽状复叶，有小叶5~8对，茎生叶托叶革质，绿色，边缘深撕裂状，下面被短柔毛或长柔毛。伞房状聚伞花序，多花，被短柔毛。萼片三角卵形，顶端渐尖或急尖。瘦果卵球形，具脉纹。花果期5~10月。

最大高度 (Max height) = 10 cm
重要值排序 (Importance value rank) = 75

Perennial herbs. Roots robust, terete. Flowering stems erect or ascending, 15-65 cm tall. Basal pinnate compound leaf with 5-8 pairs of leaflets, cauline stipules leathery, green, margin deeply lacerate, abaxially pubescent or villous. Inflorescence corymbose-cymose, many flowered. Sepals triangular-ovate, apex acute or acuminate. Petals yellow, obovate. Achenes ovoid, rugose. Fl. and fr. May-Oct..

花序　Inflorescence
摄影：林秦文　Photo by: Lin Qinwen

植株　Whole plant
摄影：林秦文　Photo by: Lin Qinwen

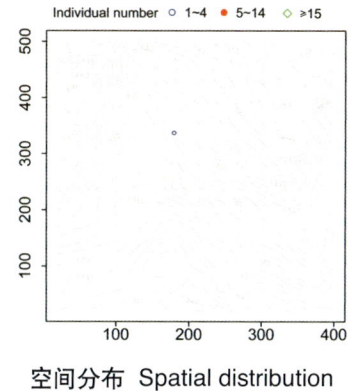

空间分布　Spatial distribution

88 等齿委陵菜

Potentilla simulatrix Th. Wolf
蔷薇科 | Rosaceae

多年生草本。纤细的匍匐茎，高15~30 cm，以及叶柄短柔毛和具长柔毛，通常具不定根在基部。叶片具3小叶；绿色小叶近无柄，位于两面，楔形倒卵形、长圆状倒卵形，或近长菱形。萼片卵状披针形，先端锐尖。花瓣黄色，倒卵形，长于萼片，先端微缺或圆形；花杜近顶生。瘦果具皱纹。花果期4~10月。

最大高度 (Max height) = 17 cm
重要值排序 (Importance value rank) = 17

Perennial herbs. Stolons slender, 15-30 cm tall, together with petioles pubescent and villous, usually with adventitious roots at base. Leaves 3-foliolate; leaflets subsessile, green on both surfaces, cuneate-obovate, oblong-obovate, or subrhomboid. Sepals ovate-lanceolate, apex acute. Petals yellow, obovate, longer than sepals, apex emarginate or rounded. Style subterminal. Achenes rugose. Fl. and fr. Apr.-Oct..

花序　Inflorescences
摄影：林秦文　Photo by: Lin Qinwen

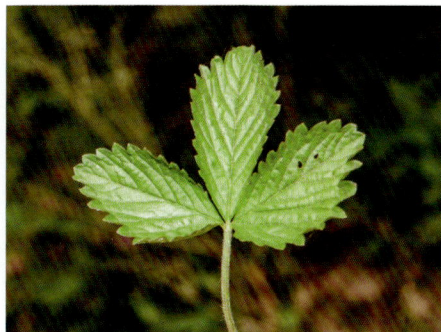

植株　Whole plant
摄影：林秦文　Photo by: Lin Qinwen

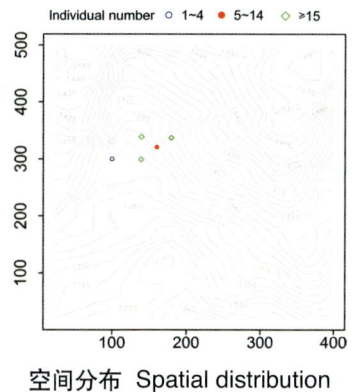

空间分布　Spatial distribution

89 葎草

lǜ cǎo | Japanese Hop

Humulus scandens (Lour.) Merr.
大麻科 | Cannabaceae

最大高度 (Max height) = 30 cm
重要值排序 (Importance value rank) = 67

多年生缠绕草本。茎、枝、叶柄均具倒钩刺。叶柄5～10 cm，叶片掌状（3～）5～7浅裂，纸质，背面在脉上具硬刺毛，正面短柔毛但不浓密，基部心形，裂片卵状三角形，具锯齿。雄花序15～25cm，黄绿色；雌花序直径约0.5 mm，苞片卵球形，纸质，具刺，先端渐尖，白色被茸毛。瘦果成熟时露出苞片外。花期春天到夏天，果期秋天。

Perennial twining herbs. Stems, branches and petioles all have barbs. Petiole 5-10 cm, leaves palmately (3-) 5-7-lobed, papery, abaxially with rigid spinulose hairs on veins, adaxially pubescent but not densely so, base cordate, lobes ovate-triangular, margin serrate. Male inflorescences 15-25 cm, male flowers, yellowish green; female inflorescences ca. 0.5 mm in diam.. Bracts ovoid, papery, spinulose, apex acuminate, white tomentose. Achenes exerted from bracts when mature. Fl. spring to summer, fr. autumn.

果序 Infructescences
摄影：林秦文 Photo by: Lin Qinwen

植株 Whole plant
摄影：刘博 Photo by: Liu Bo

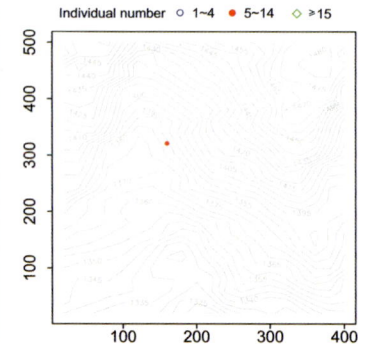

空间分布 Spatial distribution
Individual number ○ 1~4 ● 5~14 ◇ ≥15

90 狭叶荨麻

xiá yè qián má | Narrowleaf Nettle

Urtica angustifolia Fisch. ex Hornem.
荨麻科 | Urticaceae

最大高度 (Max height) = 70 cm
重要值排序 (Importance value rank) = 6

多年生草本，高40～150 cm。茎直立。茎、叶柄和叶片的两面疏生微糙硬毛，具刺毛。叶片长披针形或圆状披针形。圆锥花序状，雌雄异株；雌花基部有圆裂片，雄花花背面生有短毛及螫毛。瘦果灰褐色，广椭圆状卵形。花期7～8月，果期8～9月。

Perennial herbs, 40-150 cm tall. Stems erect. Stems, petioles, and both surfaces of leaves sparsely hirtellous and armed with stinging hairs. Leaves oblong-lanceolate to ovate-lanceolate. Inflorescences paniculate. Female flowers perianth lobes connate at base, male flowers perianth puberulent. Achene brownish gray, ovoid or broadly ovoid. Fl. Jul.-Aug., fr. Aug.-Sep..

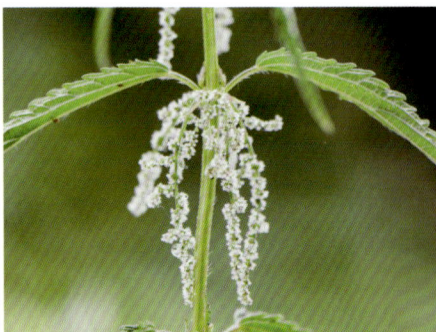

果序 Infructescences
摄影：林秦文 Photo by: Lin Qinwen

花序 Inflorescences
摄影：林秦文 Photo by: Lin Qinwen

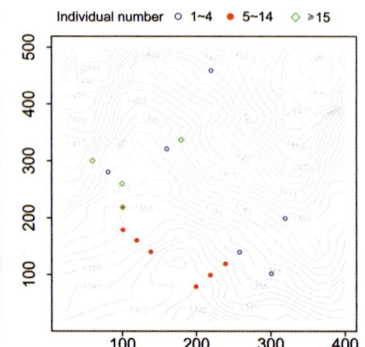

空间分布 Spatial distribution
Individual number ○ 1~4 ● 5~14 ◇ ≥15

91 裂叶堇菜

Viola dissecta Ledeb.
堇菜科 | Violaceae

多年生草本。无地上茎，植株高度变化大，花期高3～17 cm，果期高4～34 cm。基生叶，叶呈圆形、肾形或宽卵形。花淡紫色至紫堇色，上方花瓣长倒卵形，侧方花瓣长圆状倒卵形，距明显，圆筒形。蒴果长圆形或椭圆形，先端尖，果皮坚硬，无毛。花期4～9月，果期5～10月。

最大高度 (Max height) = 8.5 cm
重要值排序 (Importance value rank) = 42

Perennial herbs. Acaulescent, highly variable in height, 3-17 cm tall at anthesis, 4-34 cm tall at fruiting. Blades of basal leaves orbicular, reniform, or broadly ovate. Flowers purplish or purple-violet, upper petals narrowly obovate, lateral ones oblong-obovate, spur conspicuous, cylindric. Capsule oblong or ellipsoid, apex acute, pericarp hard, glabrous. Fl. Apr.-Sep., fr. May-Oct..

花序　　Inflorescences
摄影：林秦文　Photo by: Lin Qinwen

植株　　Whole plant
摄影：林秦文　Photo by: Lin Qinwen

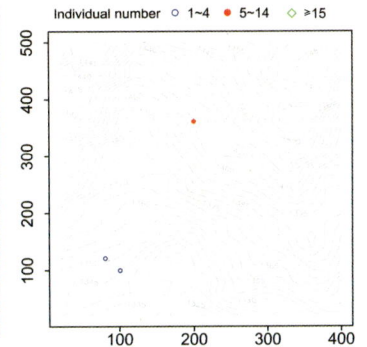

空间分布　Spatial distribution

92 斑叶堇菜

Viola variegata Fisch. ex Link
堇菜科 | Violaceae

多年生草本，高3～12 cm。根状茎通常较短而细，具数条淡褐色或近白色长根。叶均基生，呈莲座状。叶片背面通常紫红色，正面深绿色或绿色，具点条纹沿脉，圆形。花红紫色或深紫色，下部通常浅色；萼片通常略带紫色，长圆形披针形或卵状披针形；花瓣倒卵形，距圆筒状。蒴果椭圆形，无毛或疏生短毛，球状，幼时通常具短硬毛。种子褐色。花期4～8月，果期6～9月。

最大高度 (Max height) = 11 cm
重要值排序 (Importance value rank) = 85

Perennial herbs, 3-12 cm tall. Rhizome usually short, slender, with several whitish or brownish long roots. Leaves basal, rosette. Leaves abaxially usually purplish red, adaxially dark green or green, punctate-striate along veins, orbicular. Flowers red-purple or dark purple, usually light colored in lower part. Sepals usually purplish, oblong-lanceolate or ovate-lanceolate; petals obovate, spur cylindric. Capsule ellipsoid, glabrous or sparsely short hairy, globose and usually with short stiff hairs when young. Seeds brownish. Fl. Apr. -Aug., fr. Jun.-Sep..

果序　　Infructescence
摄影：林秦文　Photo by: Lin Qinwen

植株　　Whole plant
摄影：林秦文　Photo by: Lin Qinwen

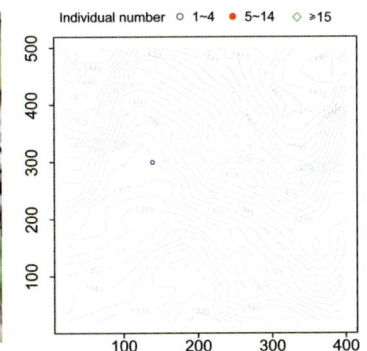

空间分布　Spatial distribution

93 鸡腿堇菜

Viola acuminata Ledeb.
堇菜科 | Violaceae

多年生草本，具地上茎。根状茎较粗壮，垂直或倾斜。2～4茎丛生，直立，高达10～40 cm。托叶大，通常羽状深裂，表面及边缘生细毛。花瓣白色、近白色或带淡紫色，侧瓣里面有须毛。蒴果。花期5～7月，果期6～9月。

最大高度 (Max height) = 18 cm
重要值排序 (Importance value rank) = 24

Perennial herbs, with aerial stems. Rhizome erect or oblique, robust. Stem usually 2-4 fasciculate, erect, 10-40 cm tall. Stipules leaflike, usually pinnatifid and fimbriate, both surfaces dotlike brown glandular. Flowers whitish violet or white, petals dotlike brown glandular, usually glabrous, rarely bearded. Capsule. Fl. May-Jun., fr. Jun.-Sep..

叶　Leaves
摄影：刘博　Photo by: Liu Bo

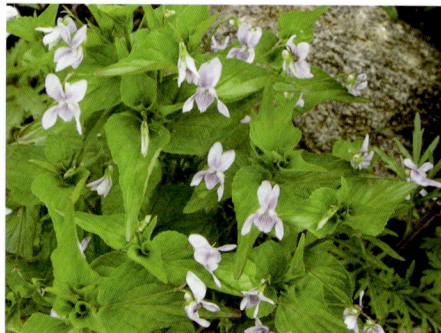

花　Flowers
摄影：林秦文　Photo by: Lin Qinwen

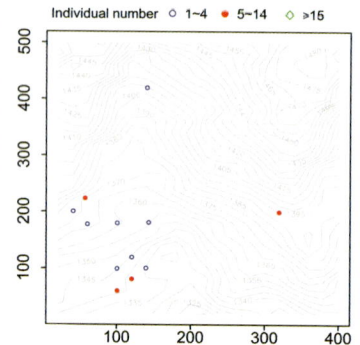

空间分布　Spatial distribution

94 北京堇菜

Viola pekinensis (Regel) W. Becker
堇菜科 | Violaceae

多年生草本，无茎，高5～9 cm。根状茎直立或斜升，1～4 cm或长，稍粗壮，具多数白色细根。叶数个，基生；叶片卵状心形，或者椭圆形。花白色，少数呈浅玫瑰色；花梗通常超过叶，细长，无毛，具2小苞片近中部。花期4～5月，果期6～8月。

最大高度 (Max height) = 5 cm
重要值排序 (Importance value rank) = 77

Perennial herbs, acaulescent, 5-9 cm tall. Rhizome erect or obliquely ascending, 1-4 cm or longer, slightly robust, with numerous white rootlets. Leaves several, basal; leaves ovate-cordate, cordate, or elliptic-cordate. Flowers white or rarely light rose; pedicels usually exceeding leaves, slender, glabrous, 2 bracteolate near middle. Fl. Apr.-May, fr. Jun.-Aug..

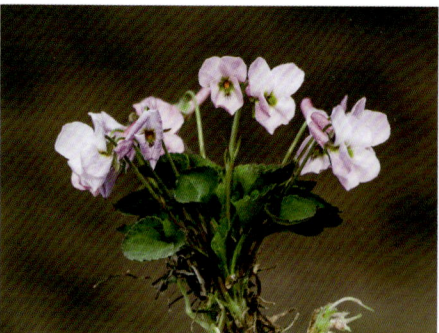

花序　Inflorescences
摄影：林秦文　Photo by: Lin Qinwen

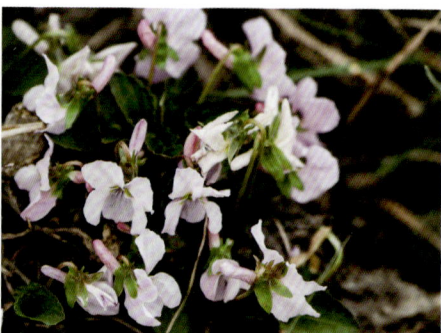

植株　Whole plant
摄影：林秦文　Photo by: Lin Qinwen

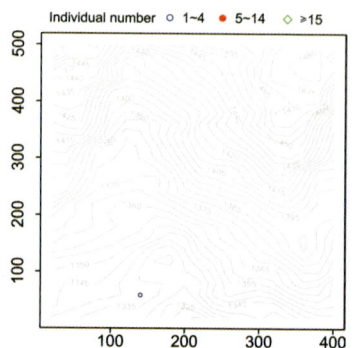

空间分布　Spatial distribution

95　毛萼堇菜

máo è jǐn cài | *Viola calyx*

Viola tenuicornis subsp. *trichosepala* W. Becker
堇菜科 | Violaceae

多年生草本。叶片被短柔毛及颗粒状凸起，叶柄具向下短毛。萼片沿边缘白色的被微柔毛的；侧花瓣明显具髯毛；子房被微柔毛。生长于向阳的山坡和空旷地带的干燥处；低于1900 m处。

最大高度 (Max height) = 4 cm
重要值排序 (Importance value rank) = 82

Perennial herbs. Leaves pubescent and granulorescent, petioles with downward pubescence. Sepals white puberulous along margin; lateral petals distinctly bearded; ovary puberulous. It grows on sunny mountain slopes, dry places in open fields; below 1900 m.

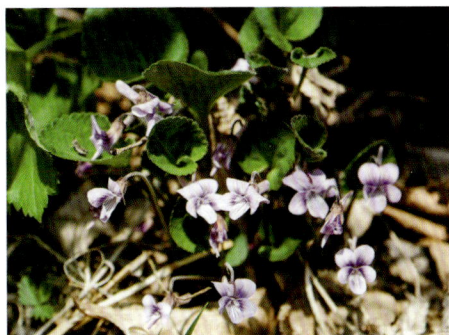

花序　Inflorescences
摄影：林秦文　Photo by: Lin Qinwen

植株　Whole plant
摄影：林秦文　Photo by: Lin Qinwen

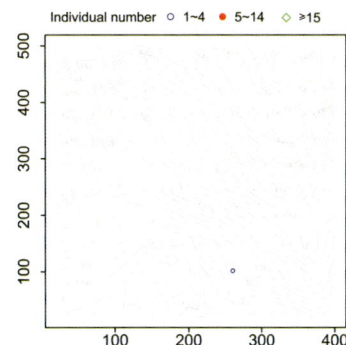

空间分布　Spatial distribution

96　球果堇菜

qiú guǒ jǐn cài | Mountain-growing Violet

Viola collina Besser
堇菜科 | Violaceae

多年生草本，高4～9 cm。叶基生，莲座状；披针形托叶，1～1.5 cm，叶片宽卵形或近圆形，基部浅或深和狭微缺，侧心形，边缘浅和钝锯齿，先端钝或锐尖，少渐尖。开花略带紫色，长约1.4 cm，长有花梗；花瓣基部呈白色。蒴果球状，密被白色柔毛；果柄成熟时通常向下弯曲。花果期5～8月。

最大高度 (Max height) = 10 cm
重要值排序 (Importance value rank) = 40

Perennial herbs, 4-9 cm tall. Leaves basal, rosulate; stipules lanceolate, 1-1.5 cm, leaves broadly ovate or suborbicular, base shallowly or deeply and narrowly sinuate, later cordate, margin shallowly and obtusely serrate, apex obtuse or acute, rarely acuminate. Flowers purplish, ca. 1.4 cm, long pedicellate, petals whitish at base. Capsule globose, densely white puberulous; fruit stalk usually curved downward when mature. Fl. and fr. May-Aug..

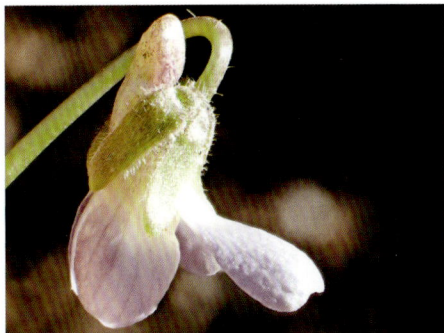

花序　Inflorescence
摄影：林秦文　Photo by: Lin Qinwen

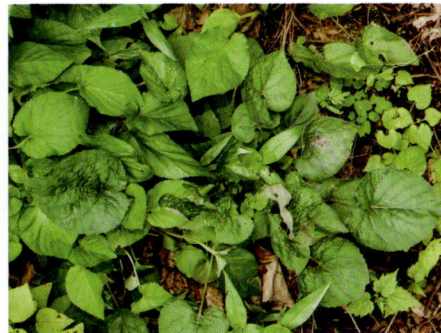

植株　Whole plant
摄影：林秦文　Photo by: Lin Qinwen

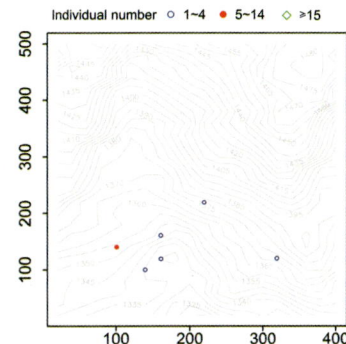

空间分布　Spatial distribution

97 毛蕊老鹳草

Geranium platyanthum Duthie
牻牛儿苗科 | Geraniaceae

最大高度 (Max height) = 24 cm
重要值排序 (Importance value rank) = 52

多年生草本，高达80 cm。叶互生，五角状肾圆形，掌状5裂，裂片菱状卵形或楔状倒卵形，上面疏被糙伏毛，下面沿脉被糙毛。花序通常为伞形聚伞花序，顶生或有时腋生；萼片长卵形或椭圆状卵形；花瓣淡紫红色，宽倒卵形或近圆形；雌蕊稍短于雄蕊，被糙毛。蒴果长约3 cm。花期6～7月，果期8～9月。

Perennial herbs, 80 cm tall. Leaves alternate, pentagonal kidney rounded, palmately 5-lobed, lobes rhomboid-ovate or wedge-obovate, proximally entire, sparsely strigose above, strigose below along veins. Cymules solitary or in aggregates at apex of each branch, sepals long ovate or elliptically ovate, petals pale purple red, broadly obovate or suborbicular, pistils slightly shorter than stamens, strigose. Capsule ca. 3 cm long. Fl. Jun.-Jul., fr. Aug.-Sep..

花序　Inflorescences
摄影：林秦文　Photo by: Lin Qinwen

果序　Inflorescences
摄影：林秦文　Photo by: Lin Qinwen

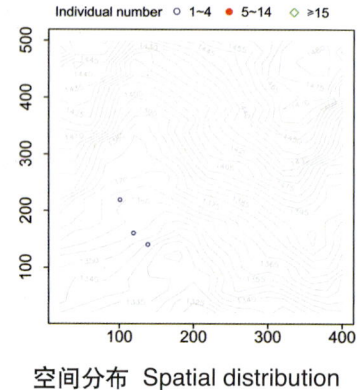
Individual number ○ 1~4 ● 5~14 ◇ ≥15
空间分布　Spatial distribution

98 高山露珠草

Circaea alpina L.
柳叶菜科 | Onagraceae

最大高度 (Max height) = 22 cm
重要值排序 (Importance value rank) = 72

草本植物，高达30 cm。茎无毛，叶卵形或宽卵形，稀圆形，基部心形或近心形，先端具牙齿。顶生总状花序，无毛或密被短腺毛。花梗无毛，呈上升状或直立；萼片椭圆形或卵形。花瓣白色，倒三角形或倒卵形。果棒状，基部平滑渐窄向果柄。花期6～8月，果期7～9月。

Herbaceous plants, 30 cm tall. Stem glabrous, leaves ovate or broadly ovate, sparsely rounded, base cordate or subcordate, apex dentate. Terminal raceme, glabrous or densely covered with short glandular hairs. Pedicels glabrous, ascending or erect. Sepals elliptic or ovate. Petals white, inverted triangular or obovate. Fruit clavate, base smooth tapering to stalk. Fl. Jun.-Aug., fr. Jul.-Sep..

花序　Inflorescences
摄影：林秦文　Photo by: Lin Qinwen

植株　Whole plant
摄影：林秦文　Photo by: Lin Qinwen

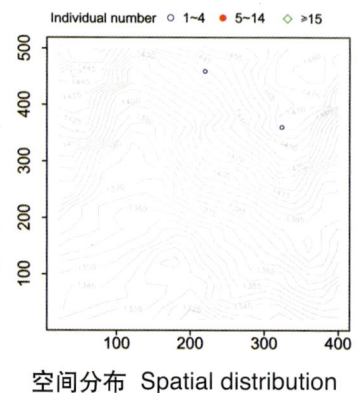
Individual number ○ 1~4 ● 5~14 ◇ ≥15
空间分布　Spatial distribution

99 白花碎米荠

bái huā suì mǐ jì | White Bittercress

Cardamine leucantha (Tausch) O. E. Schulz
十字花科 | Brassicaceae

最大高度 (Max height) = 40 cm
重要值排序 (Importance value rank) = 18

多年生草本，高25～75 cm。茎单一，不分枝，被短毛。叶为奇数羽状复叶，5小叶，小叶卵状披针形。总状花序顶生或腋生，小花12～24；花瓣白色，匙形、倒披针形到长圆形、楔形并且在基部没有爪，先端圆形。果实线形，种子近椭圆形或长圆状，褐色，狭翅或无翅。花期5～6月，果期6～7月。

Perennial herbs, 25-75 cm tall. Stems simple, flexuous, sparsely to densely villous. Cauline leaves pinnate, 5-leaflets; terminal leaflet lanceolate, elliptic, to ovate-elliptic. Racemes 12-24-flowered. Petals white, spatulate to oblong-oblanceolate, cuneate and not clawed at base, apex rounded. Fruit linear, seeds oblong, brown, narrowly winged or wingless. Fl. May-Jun., fr. Jun.-Jul..

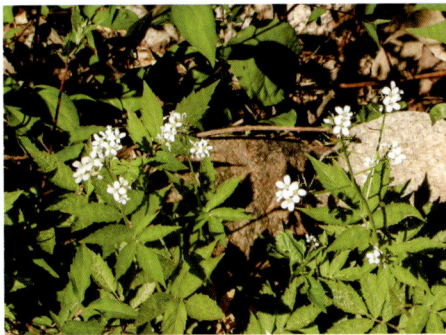

花序　Inflorescences
摄影：林秦文　Photo by: Lin Qinwen

果序　Infructescences
摄影：刘博　Photo by: Liu Bo

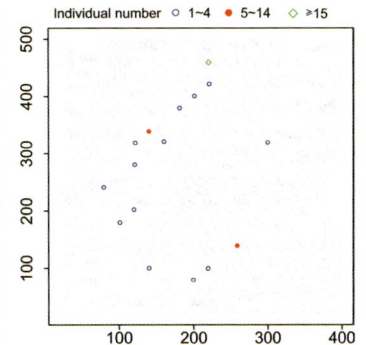

空间分布　Spatial distribution

100 糖芥

táng jiè | Bunge's Wallflower

Erysimum perofskianum Fisch. & C.A.Mey.
十字花科 | Brassicaceae

最大高度 (Max height) = 43 cm
重要值排序 (Importance value rank) = 70

多年生草本，高25～90 cm。叶片狭线形到线状披针形，灰白毛或绿色，基部渐狭，边缘全缘或具不明显小齿，先端渐尖或锐尖。具苞片的伞房状的总状花序，花密集。花瓣橙黄色，宽倒卵形或匙形，先端圆形；瓣爪离生，萼片近相等。种子长圆形。花期5～8月，果期6～10月。

Perennial herbs, 25-90 cm tall. Leaves narrowly linear to linear-lanceolate, canescent or green, base attenuate, margin entire or obscurely denticulate, apex acuminate or acute. Racemes corymbose, densely flowered. Petals orange-yellow, broadly obovate or spatulate, apex rounded; claw distinct, subequaling sepals. Seeds oblong. Fl. May-Aug., fr. Jun.-Oct..

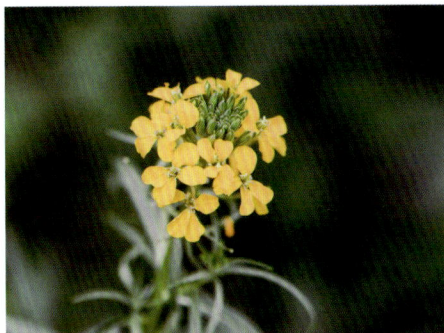

花序　Inflorescences
摄影：林秦文　Photo by: Lin Qinwen

植株　Whole plant
摄影：林秦文　Photo by: Lin Qinwen

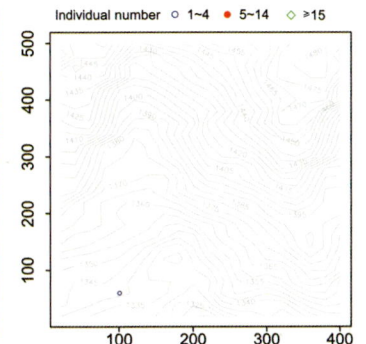

空间分布　Spatial distribution

101　浅裂剪秋罗

Silene cognata (Maxim.) H.Ohashi & H.Nakai.
石竹科 | Caryophyllaceae

最大高度 (Max height) = 36 cm
重要值排序 (Importance value rank) = 66

多年生草本，高30～90 cm。叶对生，叶片矩圆状披针形，两面被疏长毛，边缘具缘毛。二歧聚伞花序，或有时花单生在叶腋。花萼狭钟状，沿脉疏生长柔毛。花瓣橙红色，瓣片宽倒卵形，叉状浅2裂或深凹缺，裂片全缘或具不明显细齿。蒴果长卵形。花期6～7月，果期7～8月。

Perennial herbs, 30-90 cm tall. Leaves opposite, ovate-lanceolate, both surfaces sparsely hairy, margin ciliate, laxly villous at veins. Dichasium several flowered, or sometimes flowers solitary in leaf axils. Calyx narrowly campanulate, sparsely pilose along veins. Petal limb orange-red or reddish, bifid, apically obtuse, each with a subulate lateral tooth, main lobes obovate, margin entire or obscurely denticulate. Capsule long ovate. Fl. Jun.-Jul., fr. Jul.-Aug..

叶　　　　Leaves
摄影：林秦文　Photo by: Lin Qinwen

花序　　　　Inflorescences
摄影：林秦文　Photo by: Lin Qinwen

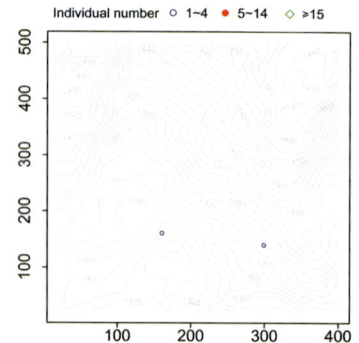

空间分布　Spatial distribution

102　水金凤

Impatiens noli-tangere L.
凤仙花科 | Balsaminaceae

最大高度 (Max height) = 78 cm
重要值排序 (Importance value rank) = 5

一年生草本，高40～80 cm。茎直立。叶互生，叶片背面绿色，正面深绿色，两面无毛，基部楔形或圆形，边缘具圆齿。花序2～4花；黄色或淡黄色，常疏生红紫色斑点。蒴果条状矩圆形，种子多数，长圆形或球形。花期6～9月，果期7～10月。

Annual herbs, 40-80 cm. Stem erect. Leaves alternate, leaves green abaxially, dark green adaxially, ovate or ovate-elliptic, both surfaces glabrous, base cuneate or rounded, margin crenate. Inflorescences 2-4-flowered; yellow or pale yellow, often sparsely with reddish purple spots. Capsule linear-cylindric, seeds many, brown, oblong-globose. Fl. Jun.-Sep., fr. Jul.-Oct..

植株　　　　Whole plant
摄影：林秦文　Photo by: Lin Qinwen

花序　　　　Inflorescence
摄影：林秦文　Photo by: Lin Qinwen

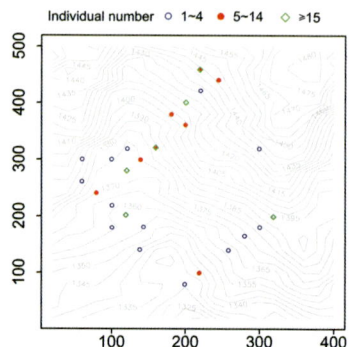

空间分布　Spatial distribution

103 茜草

Rubia cordifolia L.
茜草科 | Rubiaceae

草质攀缘藤本，茎数多条，有4棱，棱有倒生皮刺，多分枝。叶4片轮生，纸质，披针形或长圆状披针形。聚伞花序腋生或顶生；花冠淡黄或黄绿色，裂片近卵形，微伸展，无毛。浆果先橙色，成熟后变黑。花期8～9月，果期10～11月。

最大高度 (Max height) = 56 cm
重要值排序 (Importance value rank) = 14

Climbing vines herbs, stems are numerous, with 4 edges, with inverted prickles and many branches. Leaves 4-whorled, papery, lanceolate or oblong-lanceolate. Cymes axillary or termina; corolla pale yellow or greenish yellow, lobes subovate, slightly extended, glabrous. Mericarp berry becoming orange then apparently black, 4-6 mm in diam. Fl. Aug.-Sep., fr. Oct.-Nov..

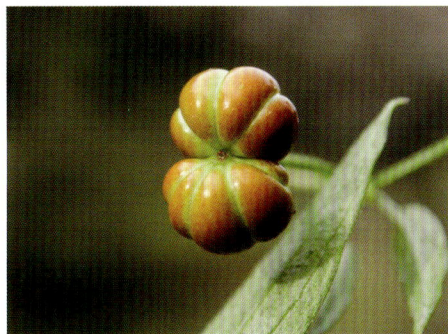

果实 Fruits
摄影：林秦文 Photo by: Lin Qinwen

植株 Whole plant
摄影：林秦文 Photo by: Lin Qinwen

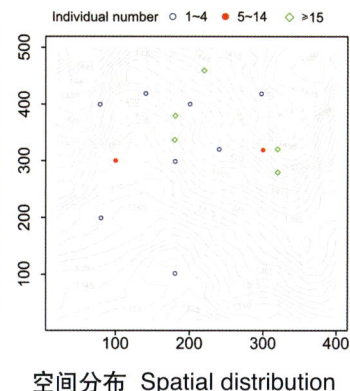

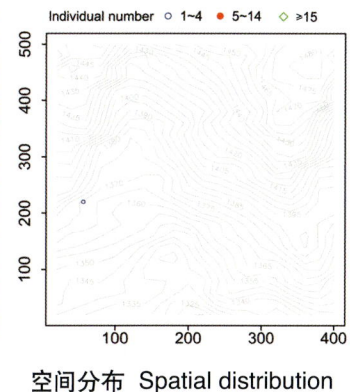

空间分布 Spatial distribution

104 林猪殃殃

Galium paradoxum Maxim.
茜草科 | Rubiaceae

多年生草本，高25 cm。有红色丝状根。茎直立，无毛或光滑。叶片膜质，近圆形，宽卵形到卵形披针形，或椭圆长圆形。聚伞花序顶生和于上部腋生，花3～11朵，常3歧分枝。花冠白色，辐状，裂片卵形。果近球形，密被黄棕色钩毛。化期5～8月，果期6～9月。

最大高度 (Max height) = 30 cm
重要值排序 (Importance value rank) = 88

Perennial herbs, 25 cm tall. Red filamentous roots. Stems erect, glabrous or smooth. Leaves membranous, suborbicular, broadly ovate to ovate-lanceolate, or elliptic-oblong. Inflorescences terminal and in axils of upper leaves with 3-11-flowered cymes, axes trichotomous and divaricate. Corolla white, rotate, lobes ovate. Fruit subglobose, densely covered with yellowish brown uncinate hairs. Fl. May-Aug., fr. Jun.-Sep..

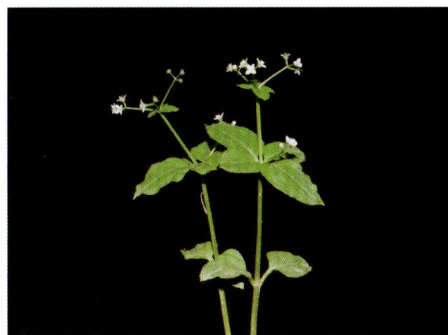

花序 Inflorescences
摄影：林秦文 Photo by: Lin Qinwen

植株 Whole plant
摄影：林秦文 Photo by: Lin Qinwen

空间分布 Spatial distribution

105 线叶拉拉藤

Galium linearifolium Turcz.
茜草科 | Rubiaceae

多年生草本。直立，基部稍木质，茎通常高65 cm，具四角棱。叶近革质，4片轮生、线形、匙形，通常稍镰刀形。聚伞花序顶生，很少腋生，疏散，少至多花，长约1.5～5 cm。花冠白色，裂片4。果椭圆状或近球状，无毛或者光滑，长2.5～3 mm。花期6～8月，果期7～9月。

最大高度 (Max height) = 25 cm
重要值排序 (Importance value rank) = 21

Perennial herbs. Erect, sometimes slightly woody at base, stems up to 65 cm tall, 4-angled. Leaves in whorls of 4, blade drying leatherylinear-spatulate, often slightly falcate. Inflorescences terminal, paniculiform, with few- to many- flowered, 1.5-5 cm long cymes. Corolla white, lobes 4. Mericarps ellipsoid to subglobose, 2.5-3 mm, glabrous and smooth. Fl. Jun.-Aug., fr. Jul.-Sep..

花序　Inflorescences
摄影：林秦文　Photo by: Lin Qinwen

植株　Whole plant
摄影：林秦文　Photo by: Lin Qinwen

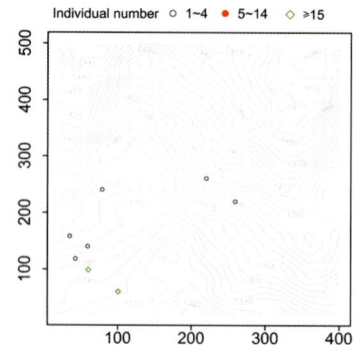

空间分布　Spatial distribution

106 香薷

Elsholtzia ciliata (Thunb.) Hyl.
唇形科 | Lamiaceae

一年生草本，高50 cm。茎无毛或具柔毛，麦秆色，老时紫褐色。叶柄0.5～3.5 cm，具狭翅；叶片卵形到椭圆形披针形，疏被细糙硬毛，正面具腺树脂疏生，下延基部楔形，边缘有锯齿，先端渐尖。穗状花序，花冠淡紫色。小坚果黄褐色，长圆形。花期7～10月，果期10月至翌年1月。

最大高度 (Max height) = 40 cm
重要值排序 (Importance value rank) = 83

Annual herbs, 50 cm tall. Stems glabrous or pilose, stramineous, purple-brown with age. Petiole 0.5-3.5 cm, narrowly winged; leaves ovate to elliptic-lanceolate, sparsely minutely hispid, adaxially sparsely resinous glandular, base cuneate decurrent, margin serrate, apex acuminate. Inflorescence spikes, corolla lavender. Nuts yellowish brown, oblong. Fl. Jul.-Oct., fr. Oct.-Jan. of next year.

花　Flowers
摄影：林秦文　Photo by: Lin Qinwen

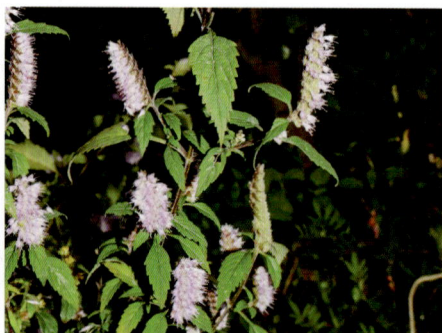

花序　Inflorescences
摄影：林秦文　Photo by: Lin Qinwen

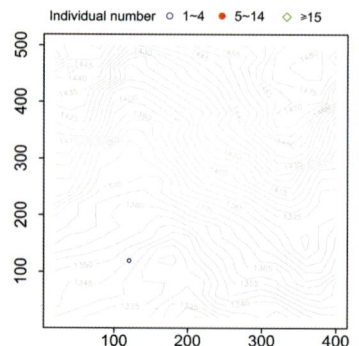

空间分布　Spatial distribution

107 蓝萼香茶菜

lán è xiāng chá cài | Japonica Isodon

Isodon japonicus var. *glaucocalyx* (Maxim.) H. W. Li
唇形科 | Lamiaceae

最大高度 (Max height) = 78 cm
重要值排序 (Importance value rank) = 84

多年生草本，高40～150 cm。根茎木质，直立，粗大，钝四棱形。茎叶对生，卵形或阔卵形。圆锥花序在茎及枝上顶生，疏松而开展。花冠淡紫、紫蓝至蓝色，上唇具深色斑点，外被短柔毛，内面无毛。小坚果卵状三棱形，长1.5 mm，黄褐色，无毛。花期7～8月，果期9～10月。

Perennial herbs, 40-150 cm tall. Rhizome woody, erect, thick, blunt quadrangular. Stem leaves opposite, ovate or broadly ovate. Panicles are terminal on stems and branches, loose and developed. Corolla is light purple, purple blue to blue, with dark spots on the upper lip, pubescent outside and glabrous inside. Nutlets ovate triangular, 1.5 mm long, yellowish brown, glabrous. Fl. Jul.-Aug., fr. Sep.-Oct..

花　　　Flowers
摄影：林秦文　Photo by: Lin Qinwen

植株　　　Whole plant
摄影：林秦文　Photo by: Lin Qinwen

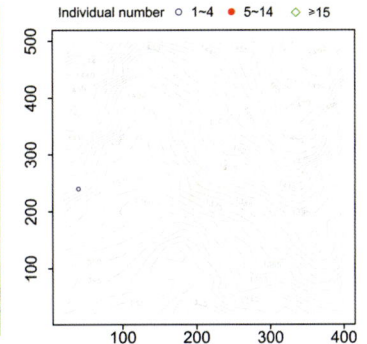

空间分布　Spatial distribution

108 黄芩

huáng qín | Baical Skullcap Root

Scutellaria baicalensis Georgi
唇形科 | Lamiaceae

最大高度 (Max height) = 37 cm
重要值排序 (Importance value rank) = 62

多年生草本。根茎肥厚，肉质，直径达2 cm，伸长而分枝。茎上升，高30～120 cm，近无毛或被上曲至开展的微柔毛。叶坚纸质，披针形至线状披针形，无毛或疏被贴生至开展的微柔毛。总状花序顶生于茎及枝上，花冠紫色、紫红至蓝色。小坚果卵球形，黑褐色。花期7～8月，果期8～9月。

Perennial herbs. Rhizomes fleshy, to 2 cm in diam., branched. Stems ascending, 30-120 cm tall, much branched, subglabrous or antrorsely to spreading puberulent. Leaves lanceolate to linear-lanceolate, papery, glab-rous or sparsely puberulent. Racemes terminal, corolla purple, purplish red to blue. Nutlets black-brown, ovoid. Fl. Jul.-Aug., fr. Aug.-Sep..

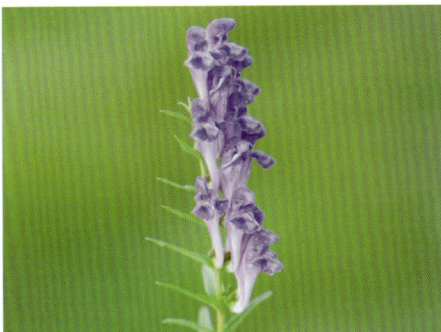

花序　　　Inflorescences
摄影：林秦文　Photo by: Lin Qinwen

植株　　　Whole plant
摄影：林秦文　Photo by: Lin Qinwen

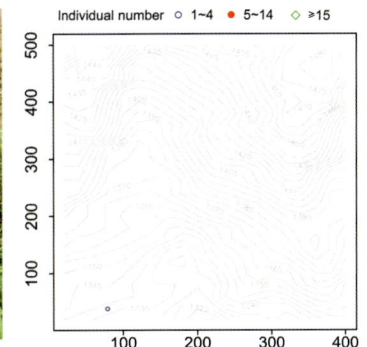

空间分布　Spatial distribution

109 京黄芩

Scutellaria pekinensis Maxim.
唇形科 | Lamiaceae

一年生草本，高24～40 cm，茎直立。叶卵圆形或三角状卵圆形。总状花序顶生。花冠蓝紫色，外被具腺小柔毛，内面无毛。冠筒前方基部略膝曲状。小坚果栗色或深棕色，卵球形，直径约1 mm，具瘤，正面伞形在基部。花期6～8月，果期7～10月。

最大高度 (Max height) = 8.5 cm
重要值排序 (Importance value rank) = 37

Annual herbs, 24-40 cm tall, stems erect. Leaves blade ovate to triangular-ovate. Racemes terminal. Corolla blue purple, glandular puberulent outside, glabrous inside. The base in front of the crown tube is slightly knee shaped. Nutlets chestnut or dark brown, ovoid, ca. 1 mm in diam., tuberculate, adaxially umbonate at base Fl. Jun.-Aug., fr. Jul.-Oct..

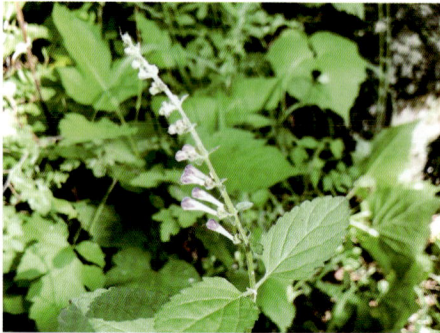

花序　Inflorescences
摄影：林秦文　Photo by: Lin Qinwen

植株　Whole plant
摄影：林秦文　Photo by: Lin Qinwen

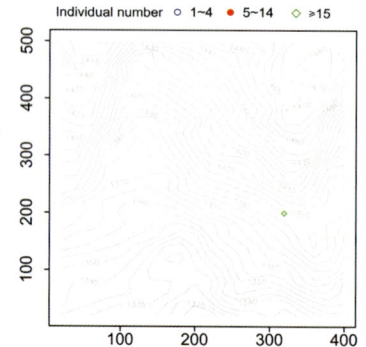

空间分布　Spatial distribution

110 糙苏

Phlomoides umbrosa (Turcz.) Kamelin & Makhm.
唇形科 | Lamiaceae

多年生草本，高达150 cm。轮伞花序通常4～8花；花萼管形；花冠淡红色或淡紫红色，下唇具少白色或红色斑点。叶圆卵形或卵状长圆形。小坚果无毛。花期6～9月，果期9月。

最大高度 (Max height) = 85 cm
重要值排序 (Importance value rank) = 8

Perennial herbs, 150 cm tall. Verticillasters 4-8-flowered; calyx tubular, corolla reddish to purple-red, rarely white with red spots on lower lip. Leaves circular-ovate to ovate-oblong. Nutlets glabrous. Fl. Jun.-Sep., fr. Sep..

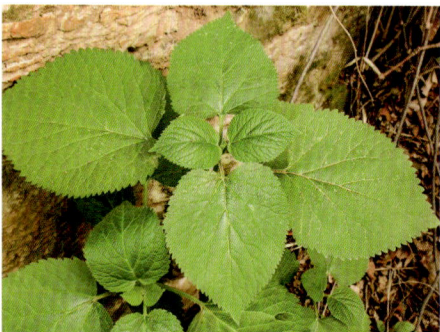

植株　Whole plant
摄影：林秦文　Photo by: Lin Qinwen

花　Flowers
摄影：林秦文　Photo by: Lin Qinwen

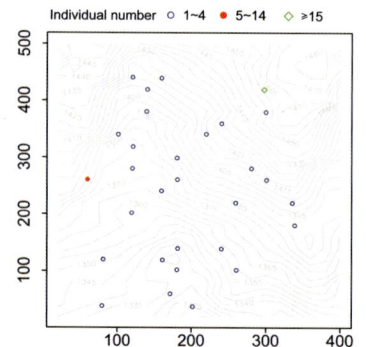

空间分布　Spatial distribution

111 菟丝子

Cuscuta chinensis Lam.
旋花科 | Convolvulaceae

一年生寄生草本。茎缠绕，黄色，纤细，直径约1 mm。花序侧生，紧密聚伞圆锥花序，很少到多花，近无柄，苞片和小苞片鳞片状。花梗约1 mm。花冠白色，瓶形，约3 mm；裂片宿存的三角状卵形，先端锐尖或钝，反折。蒴果藏于枯萎花冠，球状，直径约3 mm，周裂。种子淡褐色，卵球形，约1 mm，粗糙。

最大高度 (Max height) = 100 cm
重要值排序 (Importance value rank) = 56

Annual parasitic herbs. Stems yellow, thin, ca. 1 mm in diam.. Inflorescences lateral, compact cymose glomerules, few to many flowered, subsessile; bracts and bracteoles scalelike. Pedicel ca. 1 mm. Corolla white, urceolate, ca. 3 mm; lobes persistent triangular-ovate, apex acute or obtuse, reflexed. Capsule enclosed by withered corolla, globose, ca. 3 mm in diam., circumscissile. Seeds 1 pale brown, ovoid, ca. 1 mm, scabrous.

花序　Inflorescences
摄影：林秦文　Photo by: Lin Qinwen

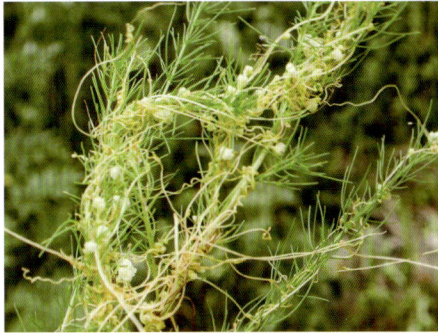

植株　Whole plant
摄影：林秦文　Photo by: Lin Qinwen

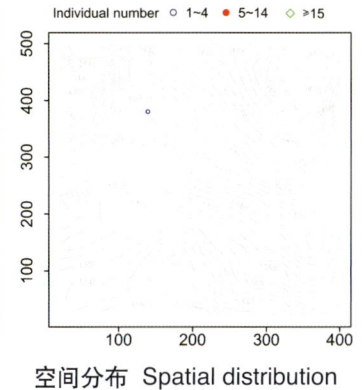

Individual number ○ 1~4　● 5~14　◇ ≥15
空间分布　Spatial distribution

112 多歧沙参

Adenophora potaninii subsp. *wawreana* (Zahlbr.) S. Ge & D. Y. Hong　桔梗科 | Campanulaceae

多年生草本。茎基常不分枝，有时有长达5 mm的分枝。茎生叶通常具叶柄，虽然有时叶柄很短；叶片变化很大，从线形到卵形，甚至在同一个体上。蒴果椭圆形。种子棕黄色，矩圆状，有一条宽棱。花期7~9月。

最大高度 (Max height) = 35 cm
重要值排序 (Importance value rank) = 49

Perennial herbs. Stem base often unbranched, sometimes with branches up to 5 mm long. Cauline leaves usually petiolate, though sometimes petiole very short; blades varying greatly, from linear to ovate, even on same individual. Capsule oval, seeds brownish yellow, rectangular, with a broad rib. Fl. Jul.-Sep..

花序　Inflorescence
摄影：姚红霞　Photo by: Yao Hongxia

植株　Whole plant
摄影：姚红霞　Photo by: Yao Hongxia

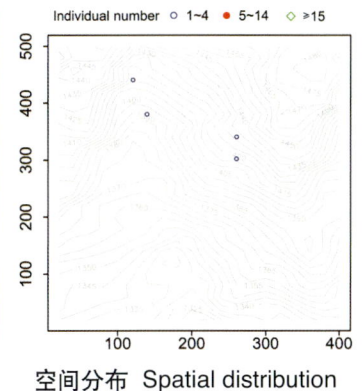

Individual number ○ 1~4　● 5~14　◇ ≥15
空间分布　Spatial distribution

113 展枝沙参 | zhǎn zhī shā shēn | Spreading Lady Bells

Adenophora divaricata Franch. & Sav.
桔梗科 | Campanulaceae

最大高度 (Max height) = 50 cm
重要值排序 (Importance value rank) = 41

叶全部轮生，很少稍交错。叶片菱形、宽椭圆形或披针形，两面无毛或具短硬毛沿脉，基部钝或楔形，边缘有粗锯齿，先端钝，锐尖。花序分枝轮生，或互生，形成一个大圆锥花序。萼裂片披针形、椭圆披针形，或线状披针形，边缘全缘；花冠蓝色或浅紫色，钟状。蒴果倒卵球形或宽椭圆形。种子金棕色，椭圆形。花期7～9月，果期9～10月。

Leaves all whorled, rarely slightly staggered. Blade rhombic, broadly elliptic or lanceolate, both surfaces glabrous or hispidulous along veins, base obtuse or cuneate, margin coarsely serrate, apex obtuse, acute, or acuminate. Inflorescence branches usually verticillate, or sometimes some alternate, forming a large panicle. Calyx lobes lanceolate, elliptic-lanceolate, margin entire. Corolla blue or light purple, campanulate. Capsule obovoid or broadly ellipsoid. Seeds golden brown, ellipsoid. Fl. Jul.-Sep., fr. Sep.-Oct..

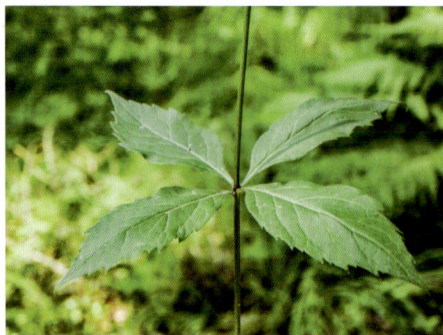

叶　　Leaves
摄影：林秦文　Photo by: Lin Qinwen

花序　　Inflorescence
摄影：林秦文　Photo by: Lin Qinwen

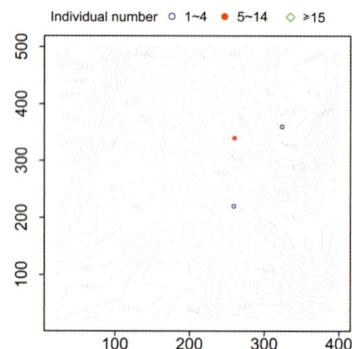

空间分布 Spatial distribution

114 石沙参 | shí shā shēn | Many-flower Lady Bells

Adenophora polyantha Nakai
桔梗科 | Campanulaceae

最大高度 (Max height) = 59 cm
重要值排序 (Importance value rank) = 48

多年生草本，高20～100 cm。茎常单生，无毛或具糙硬毛。茎生叶无梗；叶片卵形，披针形，或椭圆形，无毛或具糙硬毛，边缘具锯齿。聚伞花序具单生花形成假总状花序，或在狭窄的圆锥花序中具短花序分枝。蒴果卵球形椭圆形。种子黄棕色，卵球形椭圆形，稍压扁，约1.2 mm。花期8～10月。

Perennial herbs, 20-100 cm tall. Stem often solitary, glabrous or hispidulous. Cauline leaves sessile; blade ovate, lanceolate, or elliptic, glabrous or hispidulous, margin serrate with teeth distant. Cymes with solitary flowers forming a pseudoraceme, or in a narrow panicle with short inflorescence branches. Capsule ovoid-ellipsoid. Seeds yellow-brown, ovoid-ellipsoid, slightly compressed, ca. 1.2 mm. Fl. Aug.-Oct..

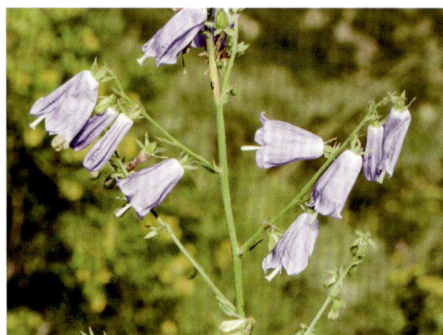

花序　　Inflorescences
摄影：林秦文　Photo by: Lin Qinwen

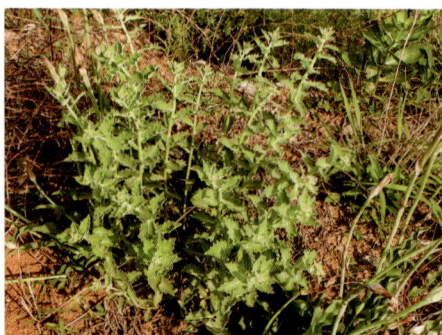

植株　　Whole plant
摄影：林秦文　Photo by: Lin Qinwen

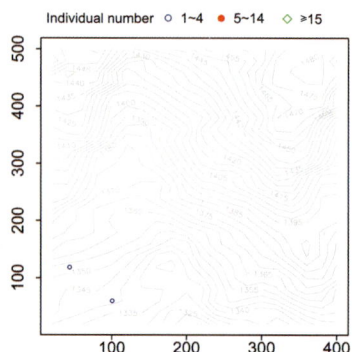

空间分布 Spatial distribution

115 和尚菜（腺梗菜）

Adenocaulon himalaicum Edgew.
菊科 | Asteraceae (Compositae)

最大高度 (Max height) = 60 cm
重要值排序 (Importance value rank) = 36

多年生草本，高30～100 cm。茎直立。叶柄10～20 cm，具翅；中部和上部叶逐渐变小，具翅叶柄；最上面的叶宽披针形，成为具苞片。叶片卵形到三角形或近圆形，边缘粗具牙齿或具小叶到具小齿或全缘。盘状头部，生在开放的圆锥状阵列中。瘦果，棍棒状至倒卵形。花期7～8月，果期8～10月。

Perennial herbs, 30-100 cm. Stem erect. Petioles 10-20 cm, winged; median and upper leaves gradually smaller, winged petiolate; uppermost leaves broadly lanceolate, becoming bracteate. Blades ovate to triangular or suborbiculate, margins coarsely dentate or lobulate to denticulate or entire. Heads disciform, borne in open paniculiform arrays. Cypselae clavate to obovoid. Fl. Jul.-Aug. fr. Aug.-Oct..

果实 Fruits
摄影：林秦文 Photo by: Lin Qinwen

植株 Whole plant
摄影：刘博 Photo by: Liu Bo

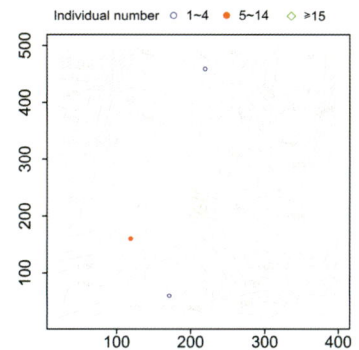

空间分布 Spatial distribution

116 苍术

Atractylodes lancea (Thunb.) DC.
菊科 | Asteraceae (Compositae)

最大高度 (Max height) = 18 cm
重要值排序 (Importance value rank) = 45

多年生草本，高30～100 cm。根状茎平卧或斜升，粗长。茎直立单生或少数茎成簇生，全部叶质地硬，硬纸质，两面同色，绿色，无毛。头状花序单生茎枝顶端，但不形成明显的花序式排列，植株有多数或少数头状花序。小花白色，长9 mm。瘦果倒卵圆状。冠毛刚毛褐色或乳白色。花果期6～10月。

Perennial herbs, 30-100 cm tall. Rhizome thick, horizontal or ascending. Stems solitary or tufted, unbranched or apically few branched. Leaves rigidly papery, green, concolorous. Inflorescences single stem branch apex, but do not form distinct inflorescence arrangement, plants have many or few inflorescences. The florets are white and 9 mm long. Achenes obovate. Crested bristles brown or white. Fl. and fr. Jun.-Oct..

花序 Inflorescences
摄影：林秦文 Photo by: Lin Qinwen

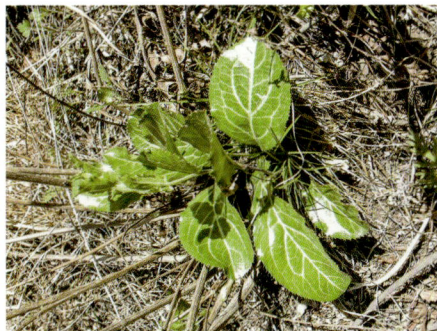

植株 Whole plant
摄影：林秦文 Photo by: Lin Qinwen

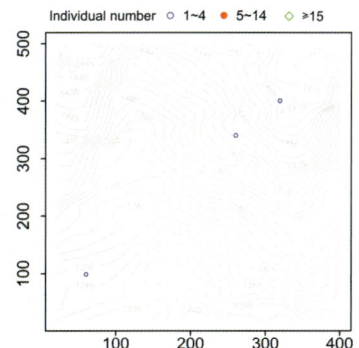

空间分布 Spatial distribution

117 山牛蒡

Synurus deltoides (Aiton) Nakai
菊科 I Asteraceae (Compositae)

最大高度 (Max height) = 20 cm
重要值排序 (Importance value rank) = 58

多年生草本，高0.7～1.5 m。根茎粗壮。茎单生，粗壮，直立，有棱。叶背面灰白色和密被毡化，正面绿色，无毛，具小刚柔毛。叶片心形、卵形、宽卵形、卵状三角形，或锯齿状，基部心形、箭形，或截形。头状花序少数。瘦果褐色，狭椭圆形，约7 mm，先端截形。冠毛刷毛棕色，1.5～2 cm。花果期6～10月。

Perennial herbs, 0.7-1.5 m tall. Rootstock stout. Stem solitary, stout, erect, ribbed. Leaves abaxially grayish white and densely felted, adaxially green, asperous, and setulose, leaves cordate, ovate, broadly ovate, ovate-triangular, or hastate, base cordate, sagittate, or truncate. Capitula several. Achene brown, narrowly ellipsoid, ca. 7 mm, apex truncate. Pappus bristles brown, 1.5-2 cm. Fl. and fr. Jun.-Oct..

果序　Infructescences
摄影：林秦文　Photo by: Lin Qinwen

植株　Whole plant
摄影：林秦文　Photo by: Lin Qinwen

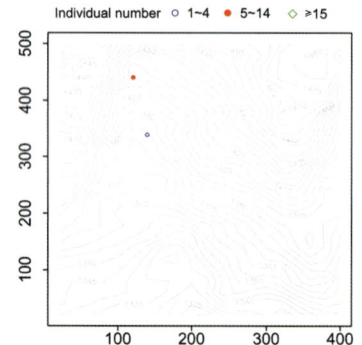

Individual number ○ 1~4 ● 5~14 ◇ ≥15

空间分布　Spatial distribution

118 银背风毛菊

Saussurea nivea Turcz.
菊科 I Asteraceae (Compositae)

最大高度 (Max height) = 50 cm
重要值排序 (Importance value rank) = 7

多年生草本，高30～120 cm。茎直立，被稀疏蛛丝毛或无毛。下部与中部茎生叶披针状三角形、心形或戟形，有锯齿。叶上面无毛，下面银灰色，密被棉毛。头状花序，有线形苞片，排成伞房状。总苞钟状，外层卵形，中层椭圆形或卵状椭圆形，内层线形。瘦果圆柱状，褐色。花果期7～9月。

Perennial herbs, 30-120 cm tall. Stems sparsely covered with spider silk hairs to glabrous, erect. Lower and middle stem leaves lanceolate triangular, cordate-shaped or halbiform, serrate. The leaves are glabrous above and silvery gray below, densely covered with cotton wool. Inflorescences head, linear bracts, arranged in corriform. Involucre campanulate, outer ovate, middle elliptic or ovate-elliptic, inner linear. Achene terete, brown. Fl. and fr. Jul.-Sep..

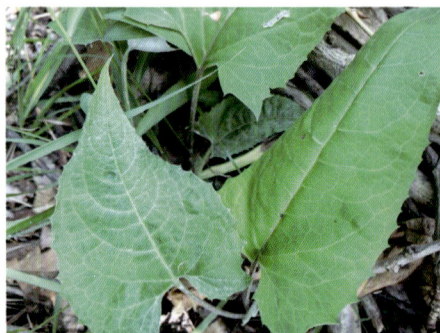

叶　Leaves
摄影：刘博　Photo by: Liu Bo

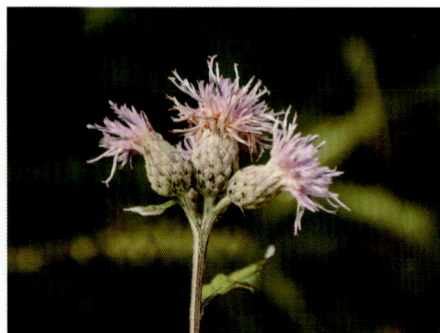

花序　Inflorescences
摄影：林秦文　Photo by: Lin Qinwen

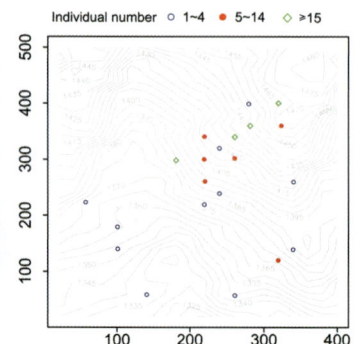

Individual number ○ 1~4 ● 5~14 ◇ ≥15

空间分布　Spatial distribution

119 蒙古风毛菊

méng gǔ fēng máo jú | Windhairdaisy

Saussurea mongolica (Franch.) Franch.
菊科 | Asteraceae (Compositae)

最大高度 (Max height) = 35 cm
重要值排序 (Importance value rank) = 2

多年生草本，高30～90 cm。茎直立，有棱，无毛或被稀疏的糙毛。叶片为卵状三角形或卵形。头状花序多数，在茎枝顶端呈伞房花序或伞房圆锥花序。总苞长圆状，总苞片5层，被稀疏的蛛丝毛或短柔毛，小花紫红色。瘦果圆柱状，褐色，无毛。冠毛2层，上部白色，下部淡褐色。花果期7～10月。

Perennial herbs, 30-90 cm. Stems erect, angulate, glabrous or sparsely hispid. The leaves are ovate triangular or ovate. Most of them are head inflorescences, which are corymbose or corymbose panicles at the top of stems and branches. Involucral bracts are oblong, with 5 layers of involucral bracts, covered with sparse spider silk hair or pubescence, the florets are purplish red. Achenes terete, brown, glabrous. 2 layers of crown hair, upper white, lower light brown. Fl. and fr. Jul.-Oct..

花序 Inflorescences	植株 Whole plant	
摄影：林秦文 Photo by: Lin Qinwen	摄影：林秦文 Photo by: Lin Qinwen	空间分布 Spatial distribution

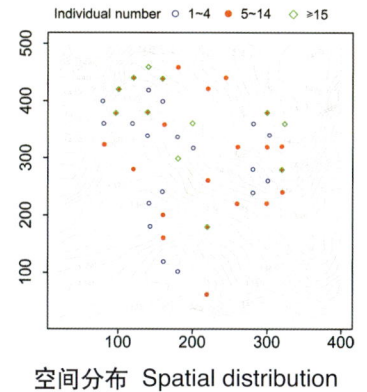

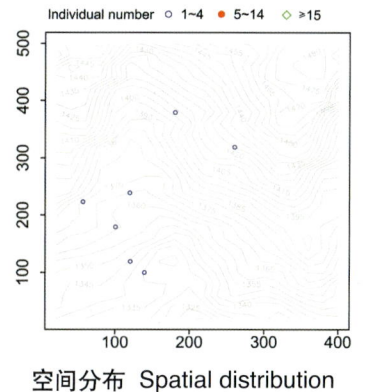

120 福王草

fú wáng cǎo | Tatarinow's Rattlesnakeroot

Nabalus tatarinowii (Maxim.) Nakai
菊科 | Asteraceae (Compositae)

最大高度 (Max height) = 25 cm
重要值排序 (Importance value rank) = 34

多年生草本，高50～150 cm。茎直立。顶裂片卵状心形、戟状心形或三角状戟形，边有不整齐的细齿，上部叶渐小，不裂。头状花序在枝上部排成圆锥花序，小花舌状，花乳白色，有时微带淡紫色，舌片顶端有5齿裂。瘦果狭长椭圆形，有5条纵肋，冠毛淡褐色。花果期8～10月。

Perennial herbs, 50-150 cm tall. Stem solitary. Terminal lobe broadly triangular-ovate, suborbicular, or broadly lanceolate in outline, margin coarsely sinuate-dentate with mucronate teeth, upper leaves tapenate, unlobed. Synflorescence paniculiform, with some to many capitula, the florets are ligulate, the flowers are milky white, sometimes microstrip lavender, the tongue tip has 5 teeth cleft. Achene narrow elliptic, with 5 longitudinal ribs. Fl. and fr. Aug.-Oct..

花序 Inflorescences	植株 Whole plant	
摄影：林秦文 Photo by: Lin Qinwen	摄影：林秦文 Photo by: Lin Qinwen	空间分布 Spatial distribution

121 多裂福王草

Nabalus tatarinowii subsp. *macranthus* (Stebbins) N. Kilian
菊科 | Asteraceae

最大高度 (Max height) = 22 cm
重要值排序 (Importance value rank) = 63

多年生草本。茎直立，上部圆锥状花序分枝，极少不分枝，全部茎枝无毛或几无毛，中下部茎叶心形或卵状心形。头状花序含5枚舌状小花，舌状小花紫色，粉红色，极少白色或黄色。瘦果线形或长椭圆状。花期7~8月，果期9~10月。

Perennial herbs. Stem erect, the upper panicle is branched, rarely unbranched, all stems and branches are glabrous or almost glabrous, and the middle and lower stems and leaves are heart-shaped or ovoid heart-shaped. The head inflorescence contains 5 tongue shaped florets, the tongue shaped florets are purple and pink, and few are white or yellow. Achenes linear or long elliptic. Fl. Jul.-Aug., fr. Sep.-Oct..

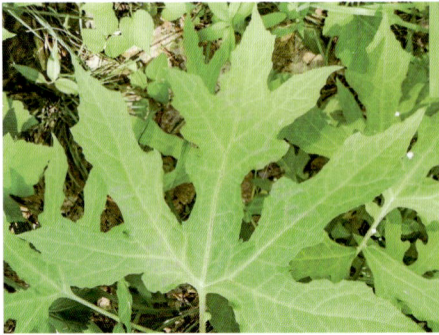

叶　Leaves
摄影：林秦文　Photo by: Lin Qinwen

植株　Whole plant
摄影：林秦文　Photo by: Lin Qinwen

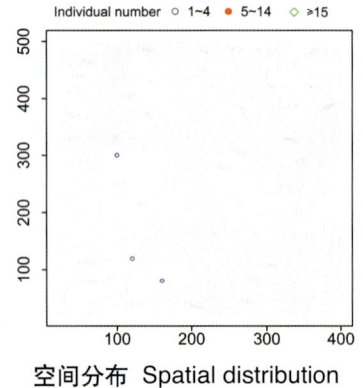

空间分布　Spatial distribution

122 山尖子

Parasenecio hastatus (L.) H. Koyama
菊科 | Asteraceae

最大高度 (Max height) = 55 cm
重要值排序 (Importance value rank) = 39

多年生草本，高40~150 cm。叶片背面淡绿色，正面绿色，基部侧裂片有时切割成小裂片，背面密被短柔毛，正面无毛或疏生微柔毛，边缘不规则细齿，先端锐尖或渐尖。瘦果淡褐色，圆筒状，6~8 mm，无毛，具棱。瘦果白色。花期7~8月，果期9~10月。

Perennial herbs, 40-150 cm tall. Blade abaxially pale green, adaxially green, basal lateral lobes sometimes incised-lobulate, abaxially densely pubescent, adaxially glabrous or sparsely puberulent, margin irregularly finely toothed, apex acute or acuminate. Achenes pale brownish, cylindric, 6-8 mm, glabrous, ribbed. Achene white. Fl. Jul.-Aug., fr. Sep.-Oct..

花序　Inflorescences
摄影：林秦文　Photo by: Lin Qinwen

叶　Leaves
摄影：林秦文　Photo by: Lin Qinwen

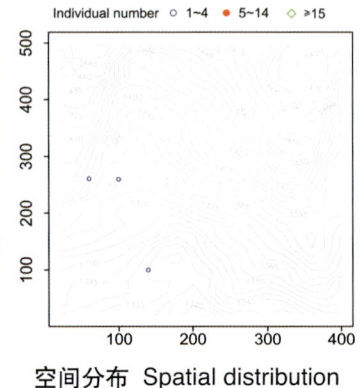

空间分布　Spatial distribution

123　三脉紫菀

Aster ageratoides Turcz.
菊科 | Asteraceae (Compositae)

多年生草本，茎高约100 cm。被柔毛或粗毛。茎下部叶宽卵圆形，中部叶窄披针形或长圆状披针形，上部叶有浅齿或全缘，叶纸质，上面被糙毛。头状花序排列成伞房状。瘦果倒卵状长圆形，灰褐色。花果期7～12月。

最大高度 (Max height) = 70 cm
重要值排序 (Importance value rank) = 3

Perennial herbs, 100 cm tall. Pilose or coarse hairs. The lower part of the stem has leaves wide ovoid, the middle part of the stem has leaves narrow lanceolate or long round lanceolate, the upper part of the stem has shallow teeth or entire leaves, leaves papery, above by rough hair. Heads disciform ,borne in open paniculiform arrys. Achene obovate oblong round, grayish brown. Fl. and fr. Jul.-Dec..

| 花序　Inflorescences 摄影：林秦文　Photo by: Lin Qinwen | 植株　Whole plant 摄影：刘博　Photo by: Liu Bo | 空间分布　Spatial distribution |

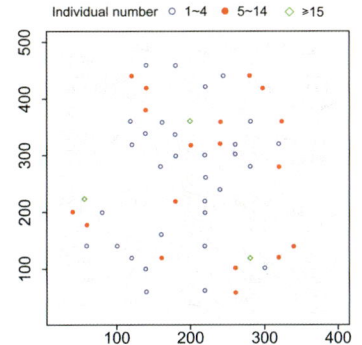

124　紫菀

Aster tataricus L. f.
菊科 | Asteraceae

多年生草本，高11～150 cm。叶片倒披针形到卵形，基部渐狭，边缘波状，粗锯齿，羽状脉6～10对。头状花序14～50或更多，在顶生伞房状合花中。瘦果深色，倒卵球形，稍压扁，2.5～3 mm，疏生到中等具糙伏毛，顶部疏生微小具柄腺。花期7～9月，果期8～10月。

最大高度 (Max height) = 15 cm
重要值排序 (Importance value rank) = 81

Perennial herbs, 11-150 cm tall. Blade oblanceolate to ovate, base attenuate, margin undulate, coarsely serrate, veins pinnately 6-10 paired. Capitula14-50 or more, in terminal corymbiform synflorescences. Achenes dark, obovoid, slightly compressed, 2.5-3 mm, sparsely to moderately strigillose, sparsely minutely stipitate glandular apically. Fl. Jul.-Sep., fr. Aug.-Oct..

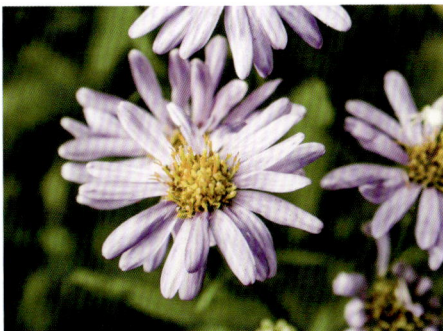

| 花序　Inflorescences 摄影：林秦文　Photo by: Lin Qinwen | 植株　Whole plant 摄影：林秦文　Photo by: Lin Qinwen | 空间分布　Spatial distribution |

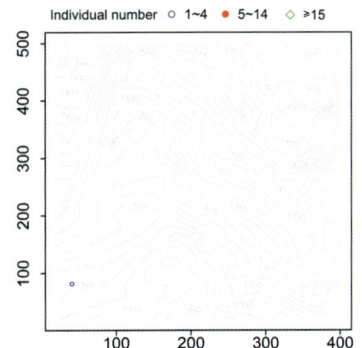

125　翠菊

Callistephus chinensis (L.) Nees
菊科 | Asteraceae

一年生或二年生草本，高20～100 cm。茎直立，单或稀疏分枝，分枝上升，有时微红，疏生到中等具长柔毛，有时疏生到中等具小柄腺体。叶下部因开花而枯萎或宿存，上部逐渐退化；叶序近等长，倒披针形，先端钝无毛。花盘小花黄色，4.5～5 mm。瘦果杂色、紫色，随年龄变灰，3～3.5 mm。花果期5～10月。

最大高度 (Max height) = 47 cm
重要值排序 (Importance value rank) = 53

Annual or biennial herbs, 20-100 cm tall. Stems erect, simple or sparingly branched, branches ascending, sometimes reddish, sparsely to moderately villous, sometimes sparsely to moderately minutely stipitate glandular. Leaves lower withered by anthesis or persistent, gradually reduced distally; phyllaries subequal, oblanceolate, apex obtuse, glabrous. Disk florets yellow, 4.5-5 mm. Achenes mottled purple, grayish with age, 3-3.5 mm. Fl. and fr. May-Oct..

花序　Inflorescences
摄影：林秦文　Photo by: Lin Qinwen

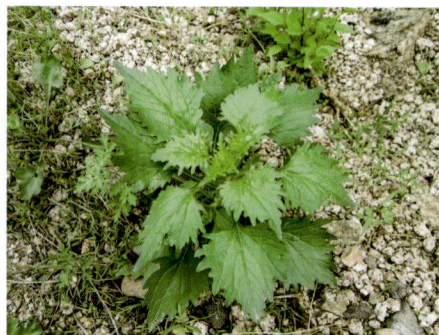

植株　Whole plant
摄影：林秦文　Photo by: Lin Qinwen

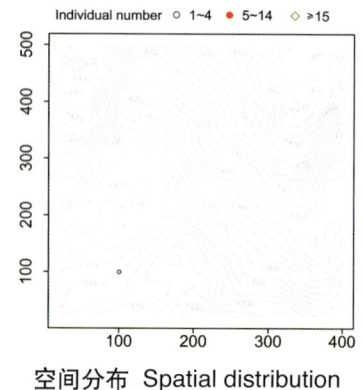

Individual number ○ 1~4 ● 5~14 ◇ ≥15

空间分布　Spatial distribution

126　小红菊

Chrysanthemum chanetii H. Lév.
菊科 | Asteraceae (Compositae)

多年生草本，高15～16 cm。茎枝疏被毛，茎中部生叶肾形、半圆形、近圆形或宽卵形，常3～5掌状或掌式羽状浅裂。上部茎叶椭圆形或长椭圆形。头状花序，少数至多数在茎枝顶端排成疏松伞房花序，舌状花为白色、粉红色或紫色。瘦果具4～6脉棱。花果期7～10月。

最大高度 (Max height) = 25 cm
重要值排序 (Importance value rank) = 12

Perennial herbs, 15-16 cm tall. Stem branches sparsely hairy, middle stem leaves reniform, semicircular, nearly round or broadly ovate, often 3-5 palmately or palmlike pinnate. Upper stem leaves elliptic or long elliptic. Inflorescences few to many, arranged in corymbose at the tips of stem branches. Ligulate flowers are white, pink, or purple. Achene with 4-6 veined ribs. Fl. and fr. Jul.-Oct..

花序　Inflorescences
摄影：林秦文　Photo by: Lin Qinwen

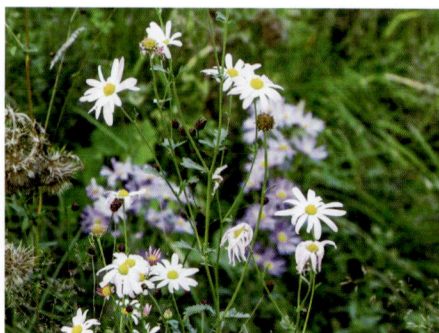

植株　Whole plant
摄影：刘博　Photo by: Liu Bo

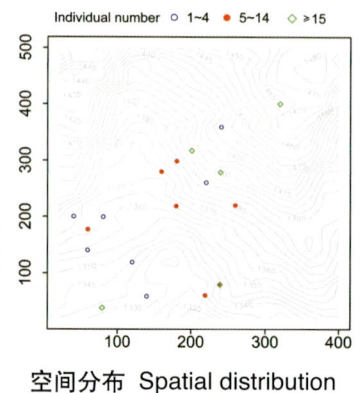

Individual number ○ 1~4 ● 5~14 ◇ ≥15

空间分布　Spatial distribution

127 甘菊

Chrysanthemum lavandulifolium (Fisch. ex Trautv.) Makino　　菊科 | Asteraceae (Compositae)

最大高度 (Max height) = 64 cm

重要值排序 (Importance value rank) = 76

多年生草本，高30～150 cm。基部和下部的茎叶因开花而枯萎。中茎叶，叶片卵形，最终裂片椭圆形。头状花序，总苞杯状，边缘白色或浅棕色干膜质，外部线形或线形长圆形，背面疏生柔毛，狭椭圆形，或倒披针形。射线花黄色。瘦果长约2 mm。花果期6～8月。

Perennial herbs, 30-150 cm tall. Basal and lower stem leaves withered by anthesis. Middle stem leaves, leaves ovate, ultimate lobes elliptic. Involucres cup-shaped, margin white or pale brown scarious, outer ones linear or linear-oblong, abaxially sparsely pilose, middle and inner ones ovate, narrowly elliptic, or oblanceolate. Ray florets yellow. achenes ca. 2 mm long. Fl. and fr. Jun.-Aug..

叶　　Leaves
摄影：林秦文　　Photo by: Lin Qinwen

植株　　Whole plant
摄影：林秦文　　Photo by: Lin Qinwen

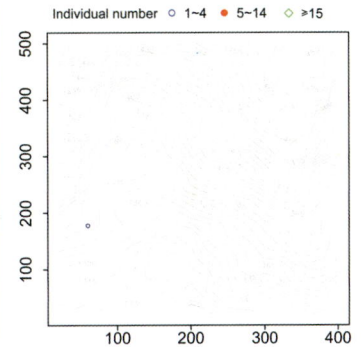

空间分布　Spatial distribution

128 野艾蒿

Artemisia lavandulifolia DC.　　菊科 | Asteraceae (Compositae)

最大高度 (Max height) = 60 cm

重要值排序 (Importance value rank) = 22

多年生草本，高50～120 cm。茎粗壮，直立，被灰白色蛛丝状柔毛。基部和最下部的叶在开花前枯萎，具长叶柄。叶片卵形或近圆形，2回羽状全裂；中间茎叶，叶片卵形，卵形椭圆形，或近圆形，背面密被绒毛，正面白色腺体斑点和疏生蛛网膜短柔毛。头状花序多数，有花序梗或无梗；总苞椭圆形或长圆形；花冠紫红色。瘦果长卵圆形或倒卵圆形。花果期8～10月。

Perennial herbs, 50-120 cm tall. Stems robust, erect, gray arachnoid pubescent. Basal and lowermost leaves withering before anthesis, long petiolate. Leaves ovate or suborbicular, 2-pinnatisect; middle stem leaves; leaves ovate, ovate-elliptic, or suborbicular, abaxially densely tomentose, adaxially white gland-dotted and sparsely arachnoid pubescent. Capitula many, pedunculate or sessile, involucre ellipsoid or oblong. corolla purple. Achenes oblong or obovoid. Fl. and fr. Aug.-Oct..

花序　　Inflorescences
摄影：林秦文　　Photo by: Lin Qinwen

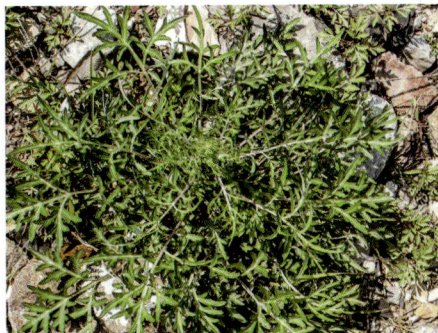

植株　　Whole plant
摄影：林秦文　　Photo by: Lin Qinwen

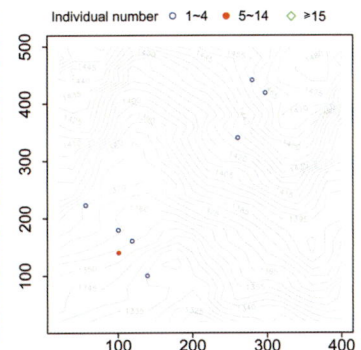

空间分布　Spatial distribution

129 蒙古蒿

Artemisia mongolica (Fisch. ex Besser) Nakai
菊科 | Asteraceae (Compositae)

最大高度 (Max height) = 65 cm
重要值排序 (Importance value rank) = 59

多年生草本，高40～120 cm，茎密被稀疏蛛丝状柔毛。中茎叶，正面密被灰白色蛛丝状短柔毛，背面被蛛丝状茸毛，叶序灰色蛛网膜短柔毛。头状花序多数，椭圆形；边缘雌花5～10；花盘小花8～15，两性。瘦果倒卵球形长圆形。花果期8～10月。

Perennial herbs, 40-120 cm tall, sparsely arachnoid pubescent. Middle stem leaves sparsely gray arachnoid pubescent adaxially, densely arachnoid tomentose abaxially. Phyllaries gray arachnoid pubescent. Inflorescences numerous, elliptic; Marginal female florets 5-10. Disk florets 8-15, bisexual. Achenes obovoid-oblong. Fl. and fr. Aug.-Oct..

花序　Inflorescences
摄影：林秦文　Photo by: Lin Qinwen

植株　Whole plant
摄影：林秦文　Photo by: Lin Qinwen

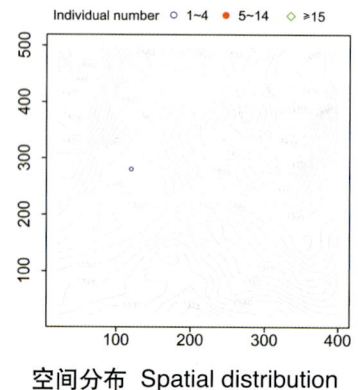

空间分布　Spatial distribution

130 南牡蒿

Artemisia eriopoda Bunge
菊科 | Asteraceae

最大高度 (Max height) = 9 cm
重要值排序 (Importance value rank) = 86

多年生草本，高80 cm。叶正面无毛，背面微被柔毛或无毛；基生叶与茎下部叶近圆形或倒卵形。头状花序宽卵圆形或近球形，直径1.5～2.5 mm，基部具线形小苞叶，排成穗状或穗状总状花序。瘦果长圆形。花果期6～11月。

Perennial herbs, 80 cm tall. Leaves adaxially glabrous, abaxially slightly pilose or glabrous. Basal leaves and lower leaves of stems are nearly round or obovate. Capitulum broadly ovoid or subglobose; 1.5-2.5 mm in diameter, with linear bracteoles at the base, arranged into spikes or spike racemes. Achenes oblong. Fl. and fr. Jun.-Nov..

花序　Inflorescences
摄影：林秦文　Photo by: Lin Qinwen

植株　Whole plant
摄影：林秦文　Photo by: Lin Qinwen

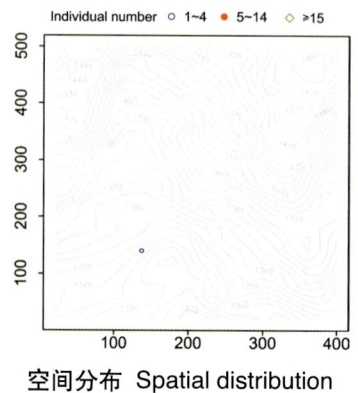

空间分布　Spatial distribution

131 阴地蒿

Artemisia sylvatica Maxim.
菊科 | Asteraceae

多年生草本，高80～130 cm。叶柄2～4 cm；叶片卵形、宽卵形、或长圆形，背面具稀疏的灰色蛛网膜被茸毛或后脱落，正面后脱落，1或2羽状深裂；裂片2或3对，椭圆形或长圆形；小叶椭圆形、椭圆状披针形，或卵状披针形，通常有锯齿。圆锥形圆锥花序；终枝纤细而柔韧；总苞近球形或卵球形。瘦果狭卵圆形或狭倒卵形。花果期8～10月。

最大高度 (Max height) = 37.5 cm
重要值排序 (Importance value rank) = 44

Perennial herbs, 80-130 cm tall. Petiole 2-4 cm; leaves ovate, broadly ovate, or oblong, abaxially sparsely gray arachnoid tomentose or glabrescent, adaxially glabrescent, 1 or 2 pinnatipartite; segments 2 or 3 pairs, elliptic or oblong; lobules elliptic, elliptic-lanceolate, or ovate-lanceolate, usually serrate. Conical panicle; ultimate branches slender and flexuous. Capitula shortly pedunculate. Achenes narrowly ovoid or narrowly obovoid. Fl. and fr. Aug.-Oct..

花序　Inflorescences
摄影：林秦文　Photo by: Lin Qinwen

植株　Whole plant
摄影：林秦文　Photo by: Lin Qinwen

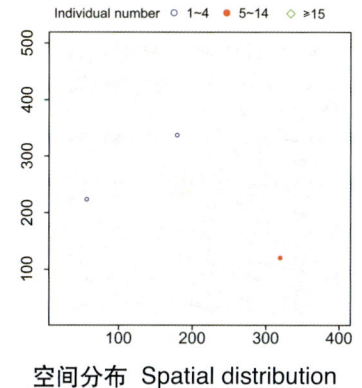

空间分布　Spatial distribution

132 歧茎蒿

Artemisia igniaria Maxim.
菊科 | Asteraceae

多年生草本，高60～120 cm。叶片卵形或宽卵形，背面密被茸毛，正面灰色被绒毛或后脱落，1或2羽状深裂；裂片2或3对，椭圆形或长圆形；小叶3或4对，先端具短尖；叶序灰色蛛网膜被茸毛。边缘雌花5～8；花盘小花7～14，两性。瘦果长圆形。花果期8～11月。

最大高度 (Max height) = 104 cm
重要值排序 (Importance value rank) = 31

Perennial herbs, 60-120 cm tall. Leaves ovate or broadly ovate, abaxially densely tomentose, adaxially gray tomentose or glabrescent, 1 or 2 pinnatipartite; segments 2 or 3 pairs, elliptic or oblong; lobules 3 or 4 pairs, apex mucronulate. Phyllaries gray arachnoid tomentose. Marginal female florets 5-8. Disk florets 7-14, bisexual. Achenes oblong. Fl. and fr. Aug.-Nov..

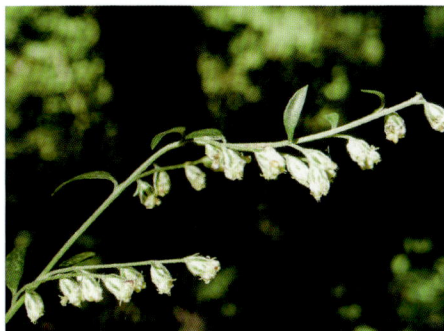

花序　Inflorescences
摄影：林秦文　Photo by: Lin Qinwen

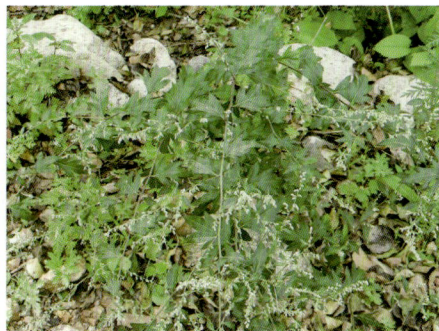

植株　Whole plant
摄影：林秦文　Photo by: Lin Qinwen

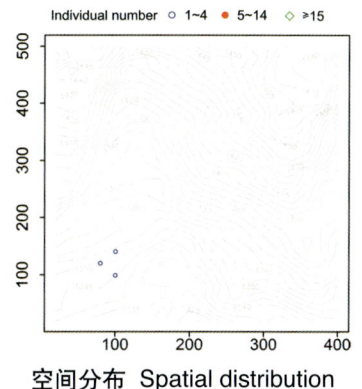

空间分布　Spatial distribution

133 黄花蒿

Artemisia annua L.
菊科 | Asteraceae

一年生草本，高70～200 cm。多分枝，疏生被微柔毛，不久无毛，强烈芳香。下部茎叶，叶片卵形或三角形卵形；中部茎叶羽状全裂。头状花序球形，径1.5～2.5 mm；雌花10～18；两性花10～30。瘦果椭圆状卵圆形，稍扁。花果期8～11月。

最大高度 (Max height) = 45 cm
重要值排序 (Importance value rank) = 65

Annual herbs, 70-200 cm tall. Much branched, sparsely puberulent, soon glabrous, strongly aromatic. Lowermost stem leaves: leaf blade ovate or triangular-ovate; middle stem leaf pinnatifid. Capitulum globose, 1.5-2.5 mm in diam.; Female flowers 10-18; Amphoteric flower 10-30. Achenes elliptic ovoid, slightly flat. Fl. and fr. Aug.-Nov..

花序　Inflorescences
摄影：林秦文　Photo by: Lin Qinwen

植株　Whole plant
摄影：林秦文　Photo by: Lin Qinwen

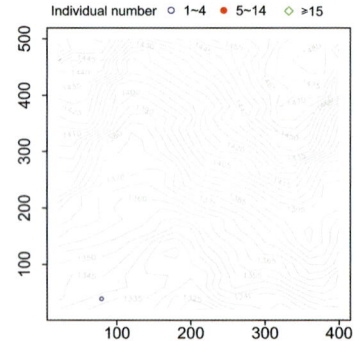

空间分布　Spatial distribution

134 五福花

Adoxa moschatellina L.
五福花科 | Adoxaceae

多年生草本，高8～15 cm。茎单一且无毛。基生叶1～3枚，为1～2回三出复叶，边缘具不整齐圆齿。茎生叶2枚，对生，三出复叶。顶生聚伞花序头状，花5～7，花绿色或黄绿色；顶生花的花萼裂片2，侧生花的花萼裂片3。核果球形。花期4～6月，果期5～7月。

最大高度 (Max height) = 5 cm
重要值排序 (Importance value rank) = 78

Perennial herbs, 8-15 cm tall. Stem simple, glabrous. Basal leaf 1-3, 1-2 ternate, leaflets obtusely lobed at margin. Cauline leaf 2, opposite, ternate. Cyme, 5-7 flowed, flower green or yellow green, terminal flower calyx 2, latera flower calyx 3. Drupe globose. Fl. Apr.-Jun., fr. May-Jul..

花　Flower
摄影：林秦文　Photo by: Lin Qinwen

植株　Whole plant
摄影：林秦文　Photo by: Lin Qinwen

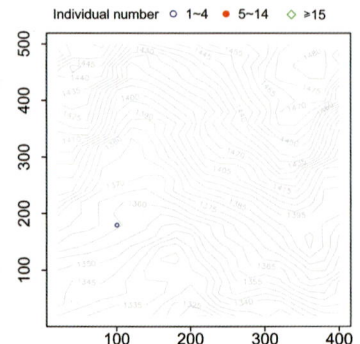

空间分布　Spatial distribution

135 异叶败酱

Patrinia heterophylla Bunge
忍冬科 | Caprifoliaceae

最大高度 (Max height) = 50 cm
重要值排序 (Importance value rank) = 74

多年生草本，高15～150 cm。茎直立，密被柔毛或近无毛。莲座状的基生叶，具叶柄，叶片狭椭圆形，羽状半裂到羽状全裂；茎生叶近无柄或具叶柄，羽状全裂的下部叶，宽卵形到线状披针形，先端渐尖到长渐尖；中部和上部叶通常不裂。花序伞房状，花冠钟状。瘦果长圆形或倒卵球形，具短硬毛。花期7～9月，果期8～10月。

Perennial herbs, 15-100 cm tall. Stems erect, densely scaberulose or subglabrous. Basal leaves rosulate, petiolate; blade narrowly elliptic, pinnatifid to pinnatisect; cauline leaves subsessile or petiolate; lower leaves pinnatisect, broadly ovate to linear-lanceolate, apex acuminate to long acuminate, middle and upper leaves often undivided. Inflorescence corymbiform, corolla campanulate. Achenes oblong or obovoid, hispidulous. Fl. Jul.-Sep., fr. Aug.-Oct..

花序　Inflorescences
摄影：林秦文　Photo by: Lin Qinwen

植株　Whole plant
摄影：林秦文　Photo by: Lin Qinwen

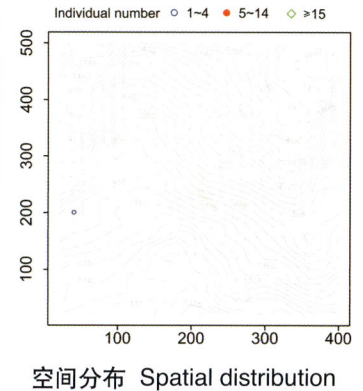

空间分布　Spatial distribution

136 北柴胡

Bupleurum chinense DC.
伞形科 | Apiaceae

最大高度 (Max height) = 80 cm
重要值排序 (Importance value rank) = 29

多年生草本，高90 cm。主根褐色。茎上部多回分枝长而开展，常呈之字曲折。基生叶披针形。复伞形花序。花瓣小舌片长圆形，顶端2浅裂。果长圆形，棕色，约3 mm × 2 mm；有突出的棱，狭翅，翅淡褐色。花期9月，果期10月。

Perennial herbs, 90 cm tall. Taproot brown. Stem upper part has many branches, often zigzag growth. Basal leaves lanceolate. Compound corymbose. Petals are small tongue shaped, oblong, and the top is 2-lobed. Fruit oblong, brown, ca. 3 mm × 2 mm; ribs prominent, narrowly wingcd, wings palc brown. Fl. Scp., fr. Oct..

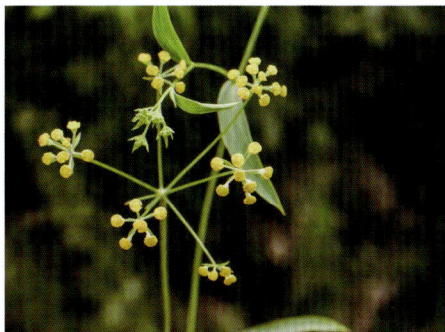

花序　Inflorescences
摄影：林秦文　Photo by: Lin Qinwen

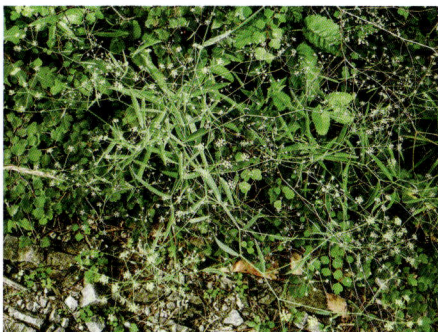

植株　Whole plant
摄影：林秦文　Photo by: Lin Qinwen

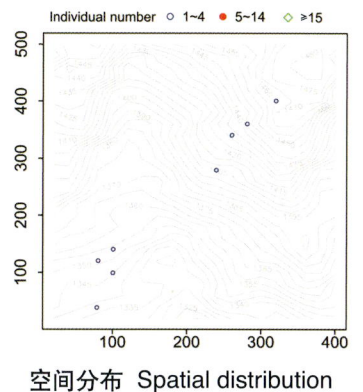

空间分布　Spatial distribution

Angelica dahurica (Fisch. ex Hoffm.) Benth. & Hook. f. ex Franch. & Sav.
伞形科 | Apiaceae

最大高度 (Max height) = 8 cm

重要值排序 (Importance value rank) = 38

多年生高大草本，高100～250 cm。根圆柱形，径3～5 cm，外表皮褐色，有浓烈气味。茎基部通常带紫色，有纵长沟纹。基生叶，有长柄，叶柄下部有管状抱茎边缘膜质的叶鞘。茎生叶2～3回羽状分裂，叶片轮廓为卵形至三角形。复伞形花序顶生或侧生。果实长圆形至卵圆形，黄棕色，有时带紫色。花期7～8月，果期8～9月。

Tall perennial herbs, 100-250 cm tall. Root cylindric, brown, 3-5 cm thick, strongly aromatic. Stem purplish green, ribbed. Basal leaves long-petiolate, sheaths oblong-inflated. Cauline leaves blade 2-3-ternate-pinnate, triangular or ovate blade. Terminal or lateral umbels. Fruit oblong to ovoid, yellowish brown, sometimes purple. Fl. Jul.-Aug., fr. Aug.-Sep..

花序　Inflorescences
摄影：林秦文　Photo by: Lin Qinwen

植株　Whole plant
摄影：林秦文　Photo by: Lin Qinwen

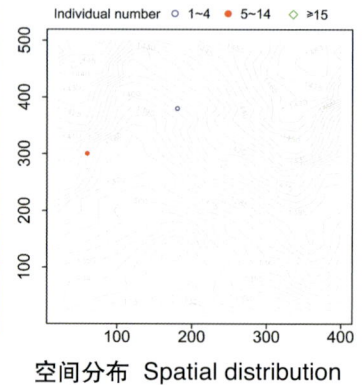

空间分布　Spatial distribution

附录I　植物中文名索引
Appendix I　Chinese Species Name Index

木本

草本

附录II　植物学名索引
Appendix II　Scientific Species Name Index

Woody Plant

Herbaceous plant